ENERGY FOR EUROPE

PRINCIPAL INTERNATIONAL OIL MOVEMENTS

1965

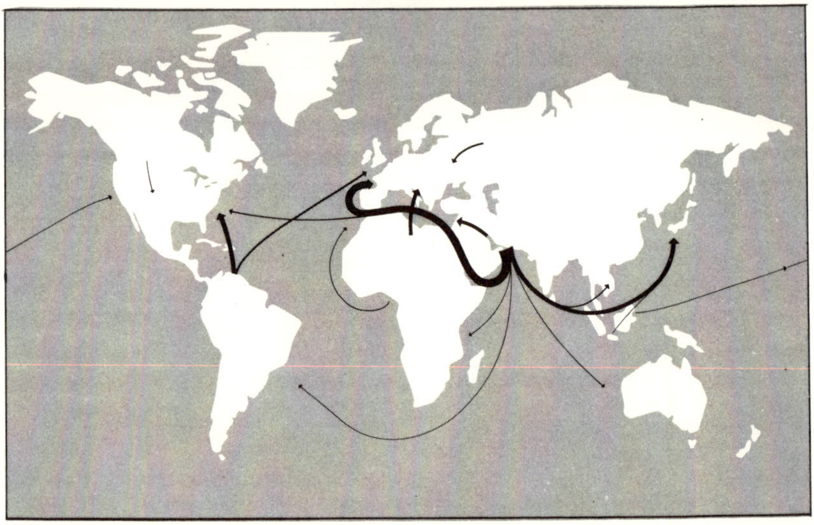

1975

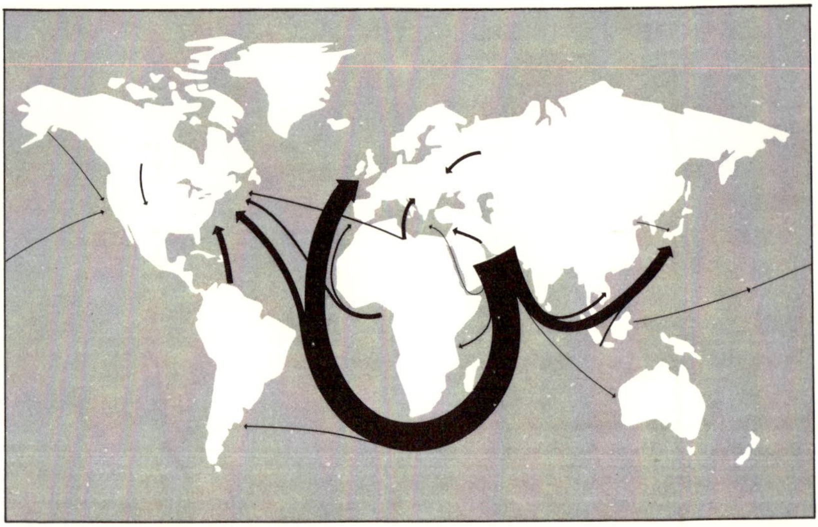

Source: Reproduced, with permission, from Exxon Background Series, *Middle East Oil,* August 1976, p. 23.

ENERGY FOR EUROPE

Economic and Political Implications

Guy de Carmoy

American Enterprise Institute for Public Policy Research
Washington, D.C.

Guy de Carmoy is a professor at the European Institute of Business Administration in Fontainebleau and at the Institute of Political Sciences in Paris.

Library of Congress Cataloging in Publication Data

Carmoy, Guy de.
Energy for Europe.

(AEI studies ; 146)
Includes bibliographical references.
1. Energy policy—Europe. 2. Power resources—Europe. I. Title. II. Series: American Enterprise Institute for Public Policy Research. AEI studies ; 146.
HD9502.E8C37 333.7 77-4289
ISBN 0-8447-3243-5

AEI studies 146

ISBN 0-8447-3243-5

Printed in the United States of America

CONTENTS

INTRODUCTION

Energy is a resource or, more precisely, a group of resources essential to all branches of economic activity and to the general well-being of man in an industrial society. Like many other problems of economic policy, energy problems require nothing more than the optimum management of limited resources.

For decades the industrial countries of Western Europe enjoyed abundant and low cost sources of energy, domestic and imported. Imported energy was under effective economic and political control. But with the end of colonialism and neocolonialism, the control of commodities—crude oil particularly—shifted from Western multinational companies to the producing states. This shift changed the whole geopolitical context. Prices soared, and the balance of payments shifted. The old quasi-mystical belief in indefinite growth and prosperity came into question. The shock in 1973–74 was much more serious to Europeans than to Americans because the former had let themselves drift into a degree of external dependence which they discovered, to their surprise, was incompatible with economic security. Energy consumption had grown unchecked while domestic production of energy had decreased since 1950. The energy gap was there to stay.

Editor's Note: In the preparation of this book I benefited greatly from the thoughtful comments of Georges Brondel, director of hydrocarbons of the European Economic Community (Brussels), of Leslie Boxer, deputy director for energy and R & D programs of the Organization for Economic Cooperation and Development (Paris), and of Dr. Judith Trunzo, of the Atlantic Institute for International Affairs (Paris), who have been kind enough to review most of the manuscript. I am also indebted to Jacques Vernant, secretary general of the Centre d'Etudes de Politique Etrangère (Paris), Louis Turner, of the Royal Institute of International Affairs (London), and, for their advice, to Professors Leslie E. Grayson and Jonathan Story, my colleagues at the European Institute of Business Administration (Fontainebleau). Mrs. Line Pierga was a most efficient and devoted secretary throughout the preparation of the manuscript.

The political disorganization of Western Europe was not conducive to the development of a common energy policy. Specialized agencies set up in the 1950s to organize a common market for coal and steel and to promote research in atomic energy had failed in their purpose, but still survived. The European Economic Community (EEC), in spite of its broader scope of activity, was unable to progress beyond the stage of recommendations. Because of the unequal distribution of domestic resources, especially after the discovery of oil and gas in the North Sea, national interests diverged. National political attitudes differed particularly toward the United States and toward the oil-producing countries of the Middle East and North Africa in respect to the Arab-Israeli conflict.

The purpose of this book is to analyze the energy situation in Western Europe at present and over the next decade. The approach is interdisciplinary, and it is made at two levels. Economic and political considerations are intermingled in the decision-making process for energy policy. Decisions are reached sometimes by an individual state, sometimes after deliberation among several states. Power politics, if not in the forefront, is always in the background.

The first chapter makes an evaluation of Western Europe's energy needs and resources in physical terms, comparing the present situation with forecasts for 1985.

The second chapter discusses the European energy situation in money terms, and includes an analysis of the effects of rising energy prices on consumption, investments, trade, and payments, and of the role of OPEC as a price-fixing cartel.

The third chapter describes the national energy policies of a number of European countries selected because of size, resources, and priorities.

The fourth chapter assesses the structure and potentialities of energy industries in Western Europe. Can these industries, fragmented into small markets, make the most of Europe's limited domestic resources?

Considerable imports of energy will be needed in the medium and long term, placing the nations of Western Europe in a position of asymmetrical interdependence in relation to their oil suppliers, especially those in the Middle East, a zone of influence contested by the United States and the Soviet Union.

The fifth chapter sets Western Europe's external supply policy in a broad geopolitical context, encompassing the strategies of the two super powers in an unstable area where Western Europe has practically no leverage.

1
NEEDS AND RESOURCES

A major difficulty in mapping an energy policy is that short- and medium-term decisions must include alternatives for long-term policies. The gap between Europe's medium-term energy needs and its resources results from a long record of complacency in the management of various sections of its energy balance sheet. At present this gap is filled by massive oil imports in an oligopolistic market structure over which the European countries have little, if any, influence. Is it possible for Europeans, in the course of the next decade, if not to bridge the gap at least to reduce it through resource substitution and energy conservation?

World Long-Term Energy Prospects

Energy is a commodity in itself and at the same time an intermediate good. It is valued as an input of some process or of utilization.[1] Industrial society, faced with an unrestricted demand for energy consumption, has enormously increased its known stock of fossil fuels, that is, of nonrenewable reserves.

The amount of recoverable reserves at any time depends on costs (which are related to interest rates), on technology (which has a bearing on costs), and on price expectations (which are related to demand). Better fields are cheaper and are developed first. Low-grade resources in distant locations with severe environmental constraints tend to price themselves out of the market.

---

[1] Kenneth Boulding, "The Economics of Energy," *Annals of the American Academy of Political Science*, November 1973.

Recoverable reserves are now estimated at some 725 billion tons of oil-equivalent, one-half of which is coal, one-third oil shale and tar sands, and only one-fifth crude oil and natural gas.[2] These reserves correspond to about 135 years of world energy consumption at the 1972 rate of 5.4 billion tons of oil-equivalent (t.o.e.). But during the 1960–72 period, the average growth rate of energy consumption was about 5 percent per annum, corresponding to a doubling every fifteen years. Assuming that such historical growth rates of consumption continued, yearly consumption would reach some 10 billion t.o.e. in 1985, 20 billion in 2000, and 40 billion in 2015, at which date the reserves, barring new discoveries, would be exhausted.

Extrapolations, however, should be used with proper caution. Those above serve to show the exponential character of the present energy-consumption pattern. Industrial countries must choose reasonable models of growth for considering cost, environment, and political balance, since they may not be able to maintain over a period of time their lion's share in the face of the claims of developing countries.

Industrial countries must also realize that the oil and natural gas era may last for a few more decades but is unlikely to continue beyond a century. On the contrary, since coal is the only fossil fuel available in large quantities, it may become once more the dominant source of energy.

In this long-term view, the prospects for Europe are rather bleak. Its coal resources are being exhausted. According to a survey that takes into account North Sea oil and gas, Europe's share of the world's recoverable reserves amounts to 2.8 percent for crude oil and 7.9 percent for natural gas.[3] At the same time, its primary energy consumption represents 20 percent of the world total.

Under the circumstances, the development of nuclear energy, despite its environmental risks, appears a necessity, unless research and development efforts can master the technology of renewable sources of energy. Solar energy is a possible answer but is not expected to be available on an industrial basis before the end of the twentieth century.

In the meantime, Europeans will have to exert themselves, after a long period of complacency, to develop a combination of energy

---

[2] U.S. Committee of the World Energy Conference, *Survey of Energy Resources*, 1974, p. 244. The survey quotes a figure of 7,600 million teracalories. One billion tons of oil is approximately equal to 10.5 million teracalories, according to the conversion tables of *BP Annual Statistical Reviews*.

[3] "World Oil and Gas Reserves," report of World Petroleum Congress, Tokyo, 1975, quoted in *The Petroleum Economist*, June 1975, p. 204.

substitution and energy conservation that will reduce their external dependence.

A Record of Complacency

The nineteenth-century industrial revolution in Europe was based on coal. Cheap coal fostered rapid industrial development in the two countries endowed with the largest exploitable reserves, Britain and Germany. Between the two World Wars, oil began to penetrate the European market, less as a competitor to coal than an addition. Oil satisfied the demand created by a new means of transportation as European auto manufacturers, following the example of America, set up assembly lines and developed mass marketing.

World War II slowed the trend, but it soon became clear that the twentieth-century industrial revolution in Europe would be based on oil.

Oil, the Dominant Industry. Since its origin, the oil industry in Europe has been characterized by a strong U.S. influence. Five out of the seven largest multinational companies are American and operate downstream facilities on a considerable scale in Europe: Exxon, Gulf, Texaco, Standard Oil of California, and Mobil. A sixth company, the Royal Dutch Shell Group, is Anglo-Dutch and ranks second in terms of sales. The seventh "major" is British Petroleum. Compagnie Française des Pétroles, the eighth in ranking, is sometimes listed among the major oil companies, with which it was associated in several Middle East ventures, sometimes with the national European companies. These include, in order of importance, ENI (Italy), Elf-Aquitaine (France), Veba-Gelsenberg (West Germany), Petrofina (Belgium), and Hispanoil (Spain).

The European companies have followed the American pattern of vertical integration and of geographical diversification of the sources of crude oil, but on a much smaller scale. In 1950, the five major American companies controlled 64 percent of Middle East reserves, or more than one-half of world recoverable reserves.[4] American political influence was dominant in the area. British influence receded with the nationalization of the local assets of the Anglo-Persian Oil Company (now B.P.) in 1951 and with the withdrawal of the British troops from the Suez Canal (1954) and Iraq (1958).

---

[4] Joseph A. Yager and Elinor B. Steinberg, *Energy and U.S. Foreign Policy* (Cambridge, Mass.: Ballinger Publishing Company, 1974), p. 10.

The major multinational companies were free to expand in the European market. Their strategy was based on the expansion of refining capacity and marketing networks and on underpricing fuel oil, the competitor of coal, while overpricing gasoline, which European car owners were eager to purchase at any price.[5] The national companies followed suit. In 1972 the major companies accounted for 58.9 percent of the refining capacity in Western Europe as against 27.2 percent for the independent companies and 13.9 percent for the state companies.[6]

In this oligopolistic market the price paid by the ultimate consumer was well above the development cost of the marginal producer. The difference constituted a rent which could be appropriated by the multinational oil companies, by the oil-consuming countries, or by the oil-producing countries. This rent was particularly high in the Middle East because of low production cost and the large quantity of reserves.[7]

In the 1950s the rent was appropriated by the major oil companies who benefited from a concessionary regime of the colonial type. Their oligopoly was shaken in the 1960s by the competition of national oil companies in France and in Italy on the one hand and of the so-called independent U.S. oil companies on the other hand. These companies obtained concessions, particularly in Libya, and granted better terms than those of the majors to governments of the producing countries. Since the American market was strongly protected from Middle Eastern and African imports by a quota system, the strategy of the majors during the 1960s was to foster the expansion of the European and Japanese markets by lowering the price of crude oil. The substitution of oil for coal had started in the late 1950s with an increase in the costs of the labor-intensive coal industry. In spite of protectionist measures taken in favor of coal, this tendency accelerated in Europe, spurred by the economic boom of the 1960s, until in 1972 Western Europe accounted for 27 percent of world oil consumption as against 20 percent in 1960.[8]

The oil-producing countries naturally resented the resulting reduction in real market prices. The Organization of Petroleum Exporting Countries (OPEC) was set up in 1960 with the purpose of dismantling the concession system. In 1971 OPEC was in a position

---

5 Romano Prodi and Alberto Clô, "Europe," *Daedalus*, Fall 1975, p. 94.

6 Christopher Tugendhat and Adrian Hamilton, *Oil, The Biggest Business* (London: Eyre Methuen, 1975), p. 296.

7 Louis Puiseux, *L'Energie et le désarroi post-industriel* (Paris: Hachette, 1973), pp. 24-25.

8 *BP Statistical Review of the World Oil Industry* (London: The British Petroleum Company Limited, 1973), p. 8.

to operate as an oligopoly through the combined effects of participation, nationalization and limitation of production. The Teheran and Tripoli agreements of 1971 confirmed the power of OPEC to control quantities and prices. The Yom Kippur war of 1973 gave the organization the opportunity to quadruple oil prices in three months, while a temporary embargo was imposed on supplies to the Netherlands, to Portugal, and to the United States. In a few years the appropriation of the rent was passed from the major oil companies to the producing countries, whose oil income thereby increased from $0.89 per barrel in October 1970 to $10.12 per barrel in January 1975 for light Arabian crude oil.[9] Thus after a brief period of sharp competition, the oligopoly of the producing countries was substituted for that of the major oil companies. The political weight of the Middle Eastern and North African states limited the influence of the United States, whose military and diplomatic activity had concentrated for almost a decade on Vietnam and whose support of Israel hampered its freedom of action in the Middle East.

Western Europe, the main consumer of Middle Eastern oil, played no part in the pricing of this commodity and assisted only as a spectator in the Arab-Israeli conflict and in the geostrategical power struggle in this zone between the two superpowers.

Oil imports in Western Europe almost quadrupled between 1960 (183.3 million tons) and 1972 (718.4 million tons). Prospecting for oil in Europe was pursued with limited success onshore in the 1960s and with greater success offshore in the North Sea in the early 1970s. But in 1972 European oil production was still very small: 19.8 million tons, or 2.7 percent of consumption.

Natural Gas, a New Contributor. While drilling in Europe, the oil companies found more gas than oil. Aside from relatively minor fields in Italy, France, and West Germany, the main discovery was the Groningen field in the Netherlands, part of the sedimental basin of the North Sea. Discoveries of natural gas followed in the southern part of the British sector and more recently in the Dutch sector of the North Sea. In 1972, production in Western Europe amounted to 150 billion cubic meters or 117.5 million t.o.e. and took care of 95 percent of domestic consumption.[10]

Natural gas is imported into Europe by pipeline from the Soviet Union and in the form of liquified natural gas (LNG) from Algeria

---

[9] Colin Robinson, *Energy Depletion and the Economics of OPEC* (Henley, England: The Henley Centre for Forecasting, 1975), Table 4, p. 19.

[10] OECD, *Energy Prospects to 1985*, vol. 1 (Paris, 1974), Table 2-5, p. 50.

and Libya. Prospecting and production of gas is carried out by oil companies granted concessions by the states, whether or not natural gas is associated with oil in the finds. Distribution of the gas is generally entrusted to a public utility company under direct control of the state or of local authorities. The natural gas industry is therefore closely linked to governments which regulate tariffs. Most national gas markets are monopolistic. The growing importance of intra- and extra-European imports gives an international character to the industry, since governments supervise all long-term import contracts. As long as the industry is purely national, it is possible to keep the price of gas lower than that of oil. But the present trend, both with European and foreign suppliers, is to link the price of gas closely to that of oil.

Coal, a Sharp Decline. The coal market in Europe has followed a pattern opposite to that of oil. Coal became more costly for two main reasons: the gradual exhaustion of the best layers in France, Belgium, and the Netherlands, and the long-term trend toward higher real wages throughout Europe, to which a labor-intensive industry was especially sensitive. French and British coal mines were nationalized after World War II, and the governments were eager to avoid labor unrest in a key industry. In 1952, the European Coal and Steel Community (ECSC) was set up among France, West Germany, Italy, and the Benelux countries to organize a common market for these two staple products. It was taken unaware by the fact that while energy consumption per capita was increasing, stocks of coal were rising from 1955 on. The ECSC authority prevented the rise in the sales price from fully reflecting the use in costs. Coal was sold at an average price and not at the marginal cost price.[11]

In 1959 the ECSC Council of Ministers refused to take formal steps to set up production and import quotas. From then on, market regulations and industrial policies were established on a national basis.

The substitution of oil for coal was the result of an oligopolistic battle between the major oil companies on the one side and the nationalized or cartelized coal companies on the other. Variable costs in the coal industry were based mainly on wages, accounting for about 60 percent of the total, and tending to rise. Variable costs in the oil industry, on the other hand, were mainly royalties and benefit taxes, in the 1960s constituting about 15 percent of the consumer price in Western Europe. Average fixed capital costs for oil tended to decrease during this period with such factors as the increase of tanker

[11] Michel Grenon, *Ce monde affamé d'énergie* (Paris: Robert Laffont, 1973), p. 150.

capacity. The result was that the price of oil, calculated on a caloric basis, was about equal to that of coal in 1955–57, and fell to about 70 percent of the price of coal in 1971.[12]

Coal gave way to oil, its share in the overall energy market falling from 61.2 percent in 1960 to 22.2 percent in 1972. Coal imports, however, increased slightly during that period, consisting mainly of coking coal from the United States and from Poland.

Nuclear Power, a Slow Start. The development of civilian nuclear power in Britain and in France began immediately after World War II, with the aid of scientists of these two countries who had been associated with the American wartime research teams. In 1954 information for the peaceful use of nuclear energy was made public, and West Germany was able to enter civilian atomic activities.

In 1957 the six countries of the European Coal and Steel Community set up a specialized organization, Euratom, to promote a European nuclear industry. But Euratom's resources dwindled as national programs expanded, and its research effort never got started.

National policies took divergent courses. Britain and France engaged in an unsophisticated technology based on natural uranium, without pooling their engineering and marketing efforts. They were unable to forestall the U.S. nuclear industry's export of a more sophisticated light water reactor (LWR), using enriched uranium, which for a long time was produced for civilian use only by the United States.[13] West Germany brought the weight of its efficient electromechanical industry on the side of the LWRs through licensing arrangements with the two main U.S. manufacturers, Westinghouse and General Electric, producing respectively the pressurized-water version and the boiling-water version.

The United States now dominates the international reactor market, and the predominance of this proven type of reactor is assured for the next fifteen years. France, Italy, and Sweden have switched from natural uranium to enriched uranium reactors. LWRs of U.S. design represented 85 percent of the nuclear power in operation in the OECD countries in 1975, and it is estimated that the proportion will rise to 89 percent by 1985.[14] Thus the two prime U.S. constructors are predominant in the world industry, directly or through foreign licens-

12 Kees Zijlstra, "The Analysis of Structural Change: Energy Supplies," paper presented at the Bruges Week Symposium, 1976, College of Europe, Bruges, Belgium, p. 3.

13 Grenon, *Ce monde affamé d'énergie*, p. 192.

14 OECD, *Uranium, Resources, Production and Demand* (Paris, 1975), Table 9, p. 29.

ing, with Westinghouse in control of by far the larger share of the European market. On the demand side, few utilities in Europe are sufficiently large to provide the flow of orders necessary for a competitive nuclear plant supply. The trend toward very large reactor and turbine generators, with their economies of scale, reinforces the imperfect nature of competition in the nuclear-plant-supply industry of the first generation.

Competition is still open in the field of advanced (as opposed to proven) reactors, the two main types being the high-temperature gas reactor (HTGR) and the liquid-metal-cooled fast-breeder reactor (LMFBR). Germany is making progress on the HTGR, and several breeder demonstration projects are operating in Europe in an advanced stage. But considering its previous achievements and the size of its domestic market, the United States is likely to keep a comparative advantage in this new field.

The needs for natural and enriched uranium are linked to the overall program of reactor construction. Europe's lower-cost, reasonably assured uranium reserves are located mostly in France; higher-cost, reasonably assured reserves are located mostly in Sweden. These amount respectively to 3.4 percent and 41 percent of world reserves.[15] Enriched uranium has been delivered to European power stations mainly by three U.S. gaseous diffusion plants operating under the aegis of the Atomic Energy Commission on the basis of long-term contracts. In 1971, when U.S. producers of enriched uranium proposed to their European clients new contracts on stringent terms with no price guarantees, two industrial groups were created to launch enrichment plants in Europe. One employed the gaseous-centrifuge technique, under French leadership, and the other, which is still in the research and development stage, is under British, Dutch, and West German sponsorship. The first European facilities will not be operational before the early 1980s.

To sum up the state of nuclear resources in Europe:

(1) U.S. manufacturers of first-generation nuclear reactors are in a position of oligopoly through direct sales and licensing in the European market.

(2) Western Europe will have to import enriched uranium mostly from the United States until it develops enrichment facilities of its own.

(3) Western Europe will have to import natural uranium for these facilities from the ore-rich countries, such as Canada, South Africa, and Australia.

---

[15] Ibid., Table 1, p. 22.

Nuclear-generated electric power accounted for only 1.4 percent of Western Europe's primary energy resources in 1972. The development of nuclear power suffered a setback in Europe in the 1960s because of the low price of crude oil and its impact on the price of coal. In 1972, nuclear-generated electricity was sometimes competitive with electricity produced from oil-fired power stations and always with electricity produced from coal-fired power stations. At a crude-oil price of $9.00 per barrel, the price relation between nuclear electricity and conventional electricity is about 1 to 1.6.[16] The cost per kilowatt of the nuclear plant varies roughly in inverse proportion to the size of the generating unit, and total cost per kilowatt hour is a function of annual station utilization. In large base-load units, nuclear-generated electricity is therefore cheaper.[17] Furthermore, the price of fuel is a small part of the production cost.

Looking back at Western Europe's overall energy balance sheet, one is struck by the gap between fast-growing demand and limited resources. Europe was influenced by a growth model based on the U.S. automotive industry without having the domestic energy resources to support it. As a result, domestic resources wound down, while imports of oil skyrocketed. European energy prices were low in the 1960s while energy prices in the United States were high, a comparative advantage which contributed to rapid European expansion. Cheap energy was taken for granted, and European governments could not conceive that major U.S. oil companies could lose their power. Security of supply was not perceived to be an issue.

The international agencies set up to deal with coal (ECSC) and with nuclear-generated electricity (Euratom) were unable to promote a common policy for their six member countries with regard to the various energy sectors, much less a comprehensive energy policy. Oil and gas fell under the jurisdiction of the European Economic Community.

The events of 1973 came as a shock to Europeans. They were not prepared for the hard times ahead.

Hard Times

From 1960 to 1985. Past performance and medium-term forecasts of Western Europe's needs and resources in primary energy are presented in Table 1. A comparison between 1960 and 1972 shows the explosion

[16] OECD, *Energy Prospects*, vol. 2, Table 9A-2, p. 185.

[17] EEC, *Guidelines for the Electricity Sector in the Community*, Com (74) 1970 (Brussels, 1974).

in energy consumption, which almost doubled in twelve years, and the increased dependence on imports, which jumped from one-third to two-thirds of the area's energy requirements.

The Organization for Economic Cooperation and Development (OECD) has made an in-depth analysis of the energy prospects of Western Europe in 1985 compared with the situation in 1972, presenting three projections. The first was based on "historical" growth, applying to 1972–85 the growth rate of the 1960–72 period, in other words, 5.5 percent. The main value of this projection is to present a favorable hypothetical situation which cannot emerge from the present predicament. OECD experts made two other alternative projections based on expectations of future oil prices, estimated respectively at $6.00 and $9.00 per barrel in 1972 prices for Arabian AP1 34° crude oil f.o.b. the Persian Gulf. These projections correspond to relative oil-price positions of $7.50 and $11.30 in 1974 prices. Because this author doubts that there will be a substantial and durable decrease in oil prices (see Chapter 2), the projection in Table 1 below is based on a price of $9.00 per barrel.

OECD experts have also assumed that the economic growth of member states will not be influenced by higher oil prices and that the previous trend of an average gross domestic product (GDP) growth rate will continue.

As explained in the OECD report, the forecasts of GDP growth rates were made before the energy crisis reached its most acute phase.[18] In the present energy situation, these forecasts may be too high for the following reasons: (a) the increase in energy prices in relation to prices for capital goods and wages will reduce energy inputs; (b) economic growth may be hindered by shifts in demand and production towards less energy-intensive products; (c) the economic slowdown registered in 1974 and 1975 will have a bearing on the overall rate of growth in the period 1974–85, in spite of the more favorable prospects for 1976.[19]

OECD states that considerable uncertainty must attach to assumptions about economic growth and the price elasticity of demand for energy. OECD assumptions may be considered a careful but somewhat optimistic assessment for the medium term in relation to economic growth, and consequently a somewhat pessimistic assess-

---

[18] OECD, *Energy Prospects*, vol. 1, chapter 1, p. 34; chapter 2, p. 41.

[19] According to OECD, *Economic Outlook*, no. 19 (Paris, July 1976), Table 2, p. 13, the volume of GDP for OECD Europe decreased by 2.3 percent in 1975 compared to 1974 and is expected to increase by 4 percent in 1976 compared to 1975.

Table 1
PRIMARY ENERGY DEMAND AND RESOURCES OF OECD EUROPE
(million t.o.e.)

	1960 [a]		1972 [b]		1985 (forecast) [b]	
	Volume	Percent	Volume	Percent	Volume	Percent
Total demand	612.0	100.0	1,157.5	100.0	2,022.7	100.0
Domestic resources	407.5	66.6	414.9	34.7	1,183.3	58.5
Net imports	204.5	33.4	755.3	65.3	839.4	41.5
Coal	21.1	3.4	31.4	2.7	52.5	2.6
Oil	183.3	30.0	718.4	57.9	666.9	33.0
Natural gas	0.1	30.0	5.5	4.7	120.0	5.9

Sources: [a] OECD, *Oil, The Present Situation and Future Prospects,* Paris, 1973. [b] OECD, *Energy Prospects to 1985,* Paris, 1974 (projection based on $9.00 per barrel at 1972 prices), vol. 1, Table 2-5, p. 50.

ment in relation to the increase in energy consumption projected in Table 1.

On the basis of a price of $9.00 per barrel, energy substitutions could reduce the European growth rate in total primary energy consumption from 5.5 percent to 4.4 percent and reduce net imports of primary sources of energy from 65.3 percent to 41.5 percent. Oil imports would be somewhat lower in 1985 (666.9 million tons) than in 1972 (718.4 million tons).

Conversely, the percentage share of the various primary sources of energy would be altered in the medium term as shown in Table 2, the shares of oil and coal being reduced and those of natural gas and especially of nuclear energy being increased. According to this projection, the share of coal would remain slightly larger than that of nuclear energy.

The external dependence of Western Europe should be considered in a geographical context, in comparison with the two major industrial countries of OECD—the United States and Japan—and in relation to Western Europe's major oil supply sources, the Middle East and Africa. Oil from these two zones provided 53 percent of Europe's total energy requirements in 1972, compared with a corresponding percentage of 59.5 for Japan and only 2.8 for the United

Table 2
PRIMARY SOURCES OF ENERGY FOR OECD EUROPE
(percentage)

	Coal Lignite	Oil	Natural Gas	Nuclear Energy	Hydroelec. Geothermal
1960 [a]	61.2	32.5	1.8	0.2	4.2
1972 [b]	22.2	63.1	10.2	1.4	3.1
1985 forecast [b]	14.8	47.8	21.5	13.5	2.4

Sources: [a] OECD, *Oil, The Present Situation and Future Prospects,* Paris, 1973. [b] OECD, *Energy Prospects to 1985,* Paris, 1974 (projection based on $9.00 per barrel), vol. 1, Tables 2-12 and 2-13, p. 57.

States. The Middle East supplied Europe with 426.4 million tons of oil and North and West Africa with 187.3 million tons. Japan imported about four-fifths of its oil from the Middle East and the bulk of the remainder from Indonesia. The two main suppliers of the United States were the Caribbean nations, with 118.8 million tons, and Canada with 53.1 million tons.

The absolute and relative growth of oil imports shows the strong leverage of the Middle Eastern and African producer countries over OECD Europe as indicated in Table 3. The United States, having become a major importer of oil, could not divert part of its production to Western Europe in 1973, as it did during two crises of supply from the Middle East—in 1956 after the Anglo-French Suez expedition and in 1967 after the Six Day war. (See frontispiece.)

Limits to Substitution. The rise in oil prices allows for energy substitutions and offers the opportunity, if not of remedying, at least of alleviating the energy imbalance within the next decade.

The first energy resources to be considered are North Sea oil and gas reserves. The estimated landed costs of North Sea crude oil (excluding royalties, taxes and profits) would amount in the mid-1980s, in 1972 prices, to $1.50-$2.00 per barrel as against a price of $0.15-$0.20 for oil in the Persian Gulf.[20] It is projected by OECD, assuming a price of $9.00 per barrel in 1972 prices, that North Sea oil production would amount to 265 million tons in 1985, or 27 percent of the oil consumption expected by that time, and that total European production including onshore would be 300 million tons.[21]

[20] OECD, *Energy Prospects,* vol. 1, Table 5-4, p. 96.
[21] Ibid., Table 5-6, p. 101.

Table 3

OIL IMPORT DEPENDENCE—WESTERN EUROPE, JAPAN, AND THE UNITED STATES

(million tons and percentage)

	Western Europe		Japan		United States	
	1960	1972	1960	1972	1960	1972
1. Total energy requirements	611.8	1,157.5	86.2	318.3	1,053.0	1,769.2
2. Total oil imports	183.3	718.4	30.9	247.5	48.0	254.6
3. 2 as percentage of 1	30.0	62.4	35.9	77.8	4.6	14.4
4. Oil imports from the Middle East and Africa	142.8	613.7	21.5	189.4	4.6	49.7
5. 4 as percentage of 1	23.3	53.0	25.0	59.5	0.5	2.8

Sources: OECD, *Oil, The Present Situation and Future Prospects*, Paris, 1973. OECD, *Energy Prospects to 1985*, Paris, 1974, vol. 1, Tables 2-2, 2-5, 2-7; and *BP Statistical Review of the World Oil Industry*, 1972.

In 1985 natural-gas production onshore (mainly in the Netherlands) and offshore (mainly in the North Sea) on the same price assumptions would reach 350 billion cubic meters or 315 million t.o.e.[22]

On a similar basis, European domestic production of coal would amount in 1985 to 246 million t.o.e., as against 230 million t.o.e. in 1972.[23] The European countries, considered as a whole, cannot hope to do more than slow down the decline of the mines, in which productivity per man-day is about one-third of the productivity in the United States.

Nuclear energy is becoming competitive with the rise in oil prices. The intrinsic advantage of nuclear power is its comparatively low fuel cost. Uranium resources are well dispersed over the OECD area and will therefore become available to Europe. The increase in nuclear fuel prices is likely to be smaller than that for fossil fuels.[24] In the past, delays in nuclear station construction were due to a number of factors, such as late delivery of equipment and public attitudes

[22] Ibid., Table 2-5, p. 50 and p. 123.

[23] Ibid., Table 2-5, p. 50.

[24] OECD, *Energy Prospects*, vol. 2, p. 183.

Table 4
SECTORAL ANALYSIS OF PRIMARY ENERGY CONSUMPTION IN OECD REGIONS—1972
(in percentages)

	OECD Europe	North America	Japan
Electricity conversion losses	14.2	13.8	12.5
Industry (including energy sector and nonenergy use)	40.5	37.3	52.2
Transportation	18.3	24.2	15.9
Residential/commercial	27.0	24.7	19.4

Source: OECD, *Energy Prospects to 1985*, Paris, 1974, vol. 1, Tables 4-1, p. 79.

regarding nuclear safety. Some uncertainties remain, casting doubts on the feasibility of the accelerated nuclear program, which calls for an installed nuclear capacity of 227 gigawatts (GWe) in 1985 as against 14.7 GWe in 1975.[25]

Such are the OECD projections for 1985. They appear ambitious. Their feasibility will be discussed later in terms of investments and best use of resources. It should be stressed at this stage that even if the proposed goals are reached by the end of the decade—at the price of hard financial, technical, and political decisions—Western Europe's dependence on external resources will remain much greater in 1985 than it was in 1960. Energy substitution is not enough: energy conservation is required.

Need for Conservation. Energy conservation consists in the reduction of energy consumption by changes in the use of energy without a comparable reduction in the gross national product and in the standard of life.

Table 4 analyzes primary energy consumption by sector in OECD Europe, North America, and Japan.

This table shows the importance of electricity conversion losses, but it does not account for the entire difference between total primary energy input and consumption of useful energy. "Overall energy efficiency can be defined as that part of all energy inputs that is finally useful to the consumer," said Kees Zijlstra in a paper presented at the Bruges Week Symposium in 1976. The author of this definition lists

[25] OECD, *Energy Prospects*, vol. 1, Table 9-2, p. 153.

four types of deductions leading to efficiencies of less than one: (1) Losses all along the transformation chain, including losses at the consumption stage. (2) Direct consumption of energy producers and transformers. (3) Indirect consumption of energy in the capital goods and durable consumption goods that ensure production, transformation, and transport of energy. (4) Direct and indirect energy consumption linked to antipollution activities in the energy field.[26]

An active energy conservation policy can produce a number of beneficial economic effects. Curtailing undue consumption in the long term postpones the depletion of domestic energy resources and in the short term diminishes the social costs of energy-generated environmental pollution. It alleviates the deficit in balance of payments resulting from higher import prices and therefore gives greater leverage to the consuming countries in their negotiations with the oil-producing countries.[27]

Higher energy prices exert a downward pressure on consumption, measured by the elasticity of demand to prices. Governments should therefore aid the proper functioning of the market by ensuring that energy prices reflect the real costs of energy. The present situation gives them an opportunity to take into account and "internalize" the environmental costs of energy consumption. The market mechanism should be supplemented by information campaigns, by revised legal norms, by financial incentives, and by other means specifically related to the main expenditures of energy, that is, in electricity conversion losses, in industry, in transportation, and in the domestic sector.

Measures for energy conservation fall into two main classes. The first reduces the demand for useful energy by asking for a sacrifice by the consumer. The second calls for a rational use of energy, reducing the energy input while placing at the disposal of the consumer the same amount of energy output. This requires improving the efficiency of equipment and processes and using energy accounting to evaluate energy consumption. Such action is linked either to the improvement of existing technologies or to the research and development of new technologies.

Methods of reducing the demand for energy are diverse. Two-thirds of the energy used for the production of electricity is dissipated in the form of heat and pollutes the environment. What is needed is

---

26 Zijlstra, "Energy Supplies," p. 10.

27 Import value of fossil fuels by EEC member countries amounted to 11.6 percent of the value of total imports in 1973 and to 20.3 percent in 1974 (Eurostat, *Basic Statistics of the Community*, 1973-74, 1974-75).

technology to combine steam generation for industry and urban heating with the production of electricity. Energy-intensive industries such as steel, aluminum, paper, plastics, and cement could provide waste heat recovery. At the same time, increased recycling of raw materials would reduce energy consumption and solid-waste disposal. Savings in space heating can be achieved by improved building design and insulation, and more efficient heating systems, such as the heat pump. To counteract the shift away from economical collective transportation by rail and bus in favor of uneconomic transportation by airlines and car, public transportation and the construction of light and durable cars should be encouraged by financial incentives. Further savings in the area of transportation could be achieved by urban planning directed toward a higher integration of residential areas and employment areas. All these techniques for saving transportation energy, however, require time. In the short term, high gasoline prices and low speed limits are the best deterrents against wasteful use of energy in individual transportation, but these measures are unpopular, and European governments are reluctant to use them.

According to the EEC Commission, the losses incurred in domestic consumption in the EEC amount to 54 percent of total needs. Some lead time, however, is required to achieve results in energy conservation. The action program prepared by the EEC Commission calls for a medium-term goal of energy savings amounting to 240 million t.o.e. in 1985, or some 15 percent of the Community's domestic consumption.[28] Such projected savings contrast with the explosion in energy consumption between 1960 and 1972. During that period the European per capita energy consumption increased from 2.548 to 4.0 tons of coal equivalent (t.c.e.).[29]

A policy of energy conservation implies a new model of growth for Europe. This new model is all the more necessary in that, as already mentioned, Europe has blindly followed the U.S. growth model without having the domestic energy resources to support it.

---

[28] Commission of the European Communities, *Rational Use of Energy* (Brussels, 1974), p. 6.

[29] *United Nations Statistical Yearbook*, 1973, Table 137, p. 350.

2
PRICES AND PAYMENTS

In Chapter 1 the energy needs and resources of Western Europe were considered in physical terms, both for the recent past and for the next decade. In this chapter Europe's energy requirements are considered in money terms. Higher oil prices affect the general pattern of supply and demand. The European countries must collect funds for investment in alternative sources of energy and in energy conservation. Substantial changes are likely to take place in the production structure as a consequence of the new allocation of resources. The European countries must finance their more expensive imports of oil. How can the resulting trade and payment deficits be met? What kind of institutions and mechanisms will help bridge the gap? And what are the prospects of a lowering of oil prices in the short or medium term, resulting from changes in the supply and demand of the various forms of energy?

Economic Effects of Higher Oil Prices

A sharp price increase in oil imports is equivalent to an excise tax, the proceeds of which are not spent in the oil-consuming countries. These proceeds are transferred to the oil-producing countries and are offset only in part by an increase in the value of exports from the oil-consuming countries to the oil-producing countries. Thus a transfer of real income takes place, which is deflationary for the oil-consuming countries.

The industrialized countries, in order to reduce their dependence on imported oil in the medium and long term, are engaging in large investment programs in energy substitution and energy conservation. These new investments are inflationary insofar as they are financed by monetary expansion and not from savings. Moreover, individuals and corporations in the consumer countries must absorb in higher oil prices a fraction of the spending power that was previously allotted to other goods and services. Either the government allows the consumption of those goods and services to fall, and thereby potentially creates unemployment, or it accepts an artificial increase in purchasing power in the form of higher nominal wages, leading to cost inflation.

In spite of increased exports, the industrialized countries are operating under a trade deficit with the oil-producing countries, so that the latter are lending directly or indirectly to their clients, while the former are building up large external debts, generally represented by short-term credits. The concentration of these credits in OPEC countries hampers the multilateral flow of capital, goods, and services and leads to a reduction in world trade, which has a multiplier effect through the reduction in demand in some export industries.

A number of structural changes have taken place in both oil-producing and oil-consuming countries as a result of the reversal of trade and the new investment patterns. Oil-producing countries are using part of their trade surplus to promote industrialization, especially in the energy-intensive industries. Oil-consuming countries are investing, as mentioned above, in energy substitution and conservation and in export industries.

To summarize, the large increase in oil prices has both deflationary and inflationary effects. It alters the flow of trade, the movement of capital, and the prospects for industrial development. These intricate economic consequences are reflected in more or less contradictory ways in the energy markets, so that forecasts of price trends must be made with caution, taking into consideration underlying political factors.

Investment Requirements

The European countries are faced with two constraints. First, they must invest in energy substitution and in energy conservation in order to reduce their external dependence. Second, they must invest in selected export industries in order to restore the equilibrium of their trade balance in the long term by satisfying the import needs of the oil-producing countries. The two policies are complementary and entail significant structural changes in internal economies.

A number of authoritative studies of the capital requirements for energy have been made by banks and international organizations.

Their estimates are based on assumptions which must continually be revised in relation to the changing economic and geopolitical picture.

OECD estimates for Western Europe during the 1974–85 period, assuming an oil price of $9.00 per barrel (in 1972 dollars), would amount to the following:

Oil and natural gas [1]	$132 billion
Nuclear electricity (accelerated program) [2]	$159 billion
Energy conservation (goal of 15 percent reduction in energy consumption) [3]	$110 billion
Total	$401 billion

These figures imply that the energy sector would take a rapidly rising share of the gross fixed capital formation. The share was 7.7 percent in the early 1970s; it would rise to 10.2 percent in 1980 and to 13.5 percent in 1985.[4]

The Commission of the European Communities has estimated capital requirements for its nine member countries representing "the differential cost to a hypothetical spontaneous trend." For the 1974–85 period these requirements in 1973 dollars would be broken down as follows:

Electricity (of which the nuclear sector amounts to $120 billion)	$180 billion
Oil and natural gas	$110 billion
Solid fuels	$ 6 billion
Total	$296 billion

Investments in various forms of energy, which amounted to about 1.5 percent of the Community product in 1965–70, would amount to 2.0 to 2.5 percent in 1974–85. Thus the additional investment would be in the range of 0.5 to 1 percent of the Community's gross domestic product.[5]

The commission made no attempt to estimate the cost of energy conservation, and the OECD report did not break down European

1 OECD, *Energy Prospects*, vol. 1, Table 5-7, p. 107.

2 OECD, *Energy Prospects*, vol. 2, Table 9A-4, p. 186.

3 Ibid., p. 85. The estimate for the whole OECD area amounts to $300 billion, with no breakdown given for Western Europe. Considering that investments in energy conservation would be proportional to the primary energy consumption in each of the subareas of OECD, the share of Western Europe would be one-third of the OECD total, or about $110 billion.

4 OECD, *Energy Prospects*, vol. 1, Table 4, p. 175.

5 EEC, *Towards a New Energy Policy Strategy for the European Community* (Brussels, 1974), p. 13.

investment requirements for electricity other than nuclear-produced. Within these limits, the estimates are consistent in the two major fields of oil-and-gas and nuclear-electricity investments. They are nonetheless subject to large margins of error because of two major factors.

First, actual investments will probably be lower in constant dollars and will not reach the ratio of 13.5 percent of the gross fixed capital formation predicted for 1985 by the OECD. The annual rate of growth in energy consumption, according to the latest returns, appears to be much lower than the forecast of 4.4 percent (see Chapter 1), and more likely to be in the 1.5 to 2 percent range.

Second, inflation has a strong impact on the sources of funding. In a 1975 study of world oil requirements and of corresponding investment needs for the 1970–85 period, the Chase Manhattan Bank stated that "it is prudent to measure the industry's financial needs in terms of at least 10 percent inflation." Under the circumstances, the demands of the oil industry in the non-Communist area would almost double in current dollars during the period under consideration.[6]

Will the energy industries, be they private companies or public utilities, be able to draw on their usual source of funds—capital markets, capital recovery, and profits—to the extent necessary? The answer will depend on the profit margins of the corporations, on the breakdown of national income between investment and consumption, and on the capability of industrial countries to control inflation.

Structural Changes

A sudden change in relative prices brings out structural changes in the economy because of the new allocation of factors of production. Higher energy prices in national economies dependent on external sources of energy lead to a shift from consumption to investment, and to different patterns of investment and consumption.

The European governments are already encouraging investment in energy substitution. They have taken steps to stop the decline in the coal industry, cancelling plans to lay off workers and close mines. Much more important has been the decision to increase exploration and production of oil and gas in the North Sea and in politically "safe" areas. The European governments are encouraging the national oil companies in this direction. On the other hand, the reduction in demand for oil in 1974 and its possible stabilization during the next

[6] The Chase Manhattan Bank, *How Much Oil? How Much Investment?* (New York, 1975).

few years has led to the postponement of some projected investments in oil refineries and in petrochemicals. These effects are short term. In the longer term, the acceleration of nuclear energy programs would give a boost to advanced technology, especially in the electrical and mechanical industries. All governments are granting credits for research in new energy sources, the lead being taken by West Germany.

In the field of energy conservation, building industries are being encouraged to invest in thermal insulation, and the public utilities which produce and distribute electricity are studying means of collecting heat manufactured as a by-product for distribution in urban areas.

The industry which offers the greatest potential for energy-saving is obviously the automobile industry. The decline in new orders for private cars has created problems for auto manufacturers in all the major European countries. From 1973 to 1974, production of passenger cars fell by 21.5 percent in West Germany, by 10.5 percent in Britain, by 9.5 percent in Italy, and by 4.4 percent in France.[7] Sales in foreign markets shrank as well. The French government was forced to grant large subsidies to Citroen and to promote a close association between that firm and Peugeot. The British government nationalized British Leyland and is subsidizing not only this domestic firm but also the Chrysler subsidiary operating in Britain. By contrast, the German government did not oppose the layoff of thousands of workers in the Volkswagen factories. On the whole, the governments have not given strong inducements to car manufacturers to expand in the area of collective, as opposed to individual, transportation. As a result, orders for new cars increased sharply in all of Western Europe in 1975 in conjunction with the rise in industrial production.

Can export-oriented industries compensate for the reduction in employment in the automobile industry that seems unavoidable in the long run? The answer to this question depends on the absorptive capacity of the oil-producing countries. Unfortunately, demand in the OPEC countries focuses on the development of industries linked with the use of cheap energy such as chemicals, petrochemicals, and construction, and of the mechanical and electrical industries. But the markets of the oil-producing countries are highly competitive, as European, Japanese, and U.S. manufacturers, with the backing of their respective governments, vie for orders for new plants. U.S. and Japanese companies have been the first to engage in a number of joint ventures, notably in Iran and Saudi Arabia. The majority of the

[7] "La Crise modifie la répartition de la production automobile dans le monde," *Le Monde*, July 8, 1975.

existing and planned ventures of the OECD multinational enterprises in the Middle East are highly capital-intensive and energy-intensive.[8] The new facilities will use the latest technologies of the industrialized countries and will enjoy the massive comparative advantage of low energy prices almost at the wellhead. The same industries in Western Europe will be at a disadvantage on these two counts. In addition, as many of the petrochemical plants under construction or planned in the Middle East are geared to export markets, risks of worldwide overcapacity in the 1980s are not to be excluded.[9]

Foreign Trade and Current Balances

To put into perspective the impact on Western Europe of the recent change in the terms of trade and especially of the rise in oil prices, one must look first at the world trade picture on the basis of the actual figures for past years and of forecasts for the future by OECD experts.

World Picture. Table 5 shows the changes in the trade balance and in the current balance of OECD, OPEC, the non-oil-producing developing countries, and other countries (mainly the Sino-Soviet area and South Africa).

The OPEC trade surplus, as could be foreseen, quadrupled from 1973 to 1974. It declined in 1975 and stood at about three-fourths of the 1974 figure in 1976. The current trade balance of OPEC follows somewhat the same pattern, the surplus declining from $65 billion in 1974 to $40 billion in 1975. The reduction in the OPEC surplus is due on the one hand to the increased import capacity of this group of countries (see below) and on the other hand to the decline in volume of OECD oil imports. This decline was estimated at 4 percent in 1974 compared with 1973 [10] and at 8 percent in 1975 compared with 1974.[11]

On the OECD side, the trade balance ran a $7.5 billion surplus in 1973, a $26.5 billion deficit in 1974, and in 1975 registered a surplus almost equal to that of 1973. Figures for the current balance are somewhat lower than those of the trade balance, with a small $5.5 bil-

---

[8] Laurence G. Franko, *Prospects for Industrial Joint Ventures in the Oil Exporting Countries of the Middle East and North Africa* (Paris: OECD Development Center, 1975), p. 719.

[9] "Concern over Middle East Potential," *The Petroleum Economist*, August 1976, p. 301.

[10] OECD, *Economic Outlook*, no. 17 (Paris, June 1975), p. 79.

[11] OECD, *Economic Outlook*, no. 18 (Paris, December 1973), p. 58.

Table 5

WORLD TRADE AND CURRENT BALANCES
(in billions of dollars)

	1973	1974	1975	1976 (est.)	1977 (forecast)
Trade balances					
OECD	7.5	−26.5	5.5	−13.0	−5.0
OPEC	19.5	82.5	52.0	61.0	57.0
Non-oil-producing developing countries	−2.5	−21.0	−30.0	−21.0	−23.0
Other	−6.0	−7.0	−14.5	−12.5	−13.5
Discrepancy	18.5	28.0	13.0	14.5	15.5
Current balances (including official transfers)					
OECD	2.5	−33.0	−6.5	−22.5	−17.5
OPEC	3.5	65.5	34.5	42.0	36.5
Non-oil-producing developing countries	−2.5	−21.5	−32.5	−24.0	−26.5
Other[a]	−4.0	−6.5	−14.0	−12.5	−14.5
Discrepancy	−0.5	4.5	−18.0	−17.0	−22.0

[a] Sino-Soviet area, South Africa, Israel, Cyprus, Malta and Yugoslavia.

Source: OECD, *Economic Outlook*, Paris, no. 18, December 1975, Table 29, p. 64; no. 19, July 1976, Table 28, p. 55; no. 20, December 1976, Table 24, p. 67.

lion deficit for 1975 and a much larger one for 1976. These global results include important regional differences as explained below, but apparently OECD as a whole has weathered the shock of the oil crisis.

Such is not the case for the non-oil-producing developing countries, whose trade and current balances were almost in equilibrium in 1973, and who registered trade and current-account deficits in the order of $16 billion in 1974 and of $24.5 billion in 1975, with only a slight improvement estimated for 1976. This group of countries has been the main victim of the oil crisis. The situation will eventually boomerang on world trade, as non-oil-producing developing countries will not be in a position to maintain their level of imports from the industrialized countries and especially from Western Europe.

OECD and the European Picture. Table 6 shows the changes in the trade balances and in the current balances of the United States, Japan, and the European countries of OECD, the latter being broken down into three groups, the EEC, northern Europe, and southern Europe.

Table 6

OECD AND WESTERN EUROPE TRADE BALANCES AND CURRENT BALANCES
(in billions of dollars)

	Trade Balances					Current Balances				
	1973	1974	1975	1976 (est.)	1977 (forecast)	1973	1974	1975	1976 (est.)	1977 (forecast)
United States	1.0	−5.3	9.0	−8.5	−10.5	0.3	−0.6	11.7	−1.25	−3.0
Japan	3.7	1.4	5.0	10.25	7.5	−0.1	−4.7	−0.7	3.75	0.0
EEC	6.5	−4.5	9.6	0.0	8.0	0.5	−12.25	0.7	−8.25	1.0
France	0.8	−3.9	1.4	−4.25	−2.0	−0.7	−6.0	−0.1	−6.0	−3.75
West Germany	15.0	21.9	17.2	16.5	19.25	4.3	9.6	3.9	4.0	5.0
Italy	−4.0	−8.5	−1.2	−3.25	0.0	−2.7	−7.8	−0.6	−2.75	0.5
United Kingdom	−5.6	−12.2	−7.1	−6.75	−5.0	−2.9	−8.8	−3.7	−3.25	−1.25
Belgium/ Luxembourg	1.4	0.6	0.1	−0.75	−0.5	1.3	0.7	0.3	−0.25	−0.25
Netherlands	0.7	0.2	0.9	1.75	2.25	1.8	1.6	1.6	2.0	2.75
Denmark	−1.6	−1.8	−1.3	−2.5	−2.25	−0.5	−0.9	−0.5	−1.75	−1.5
Ireland	0.0	−1.0	−0.4	−0.5	−0.75	−0.2	−0.7	0.0	−0.25	−0.25
Northern Europe[a]	−2.6	−5.9	−4.8	−4.0	−2.5	0.2	−3.7	−4.2	−3.5	−2.5
Southern Europe[b]	−7.2	−13.3	−14.1	−14.0	−15.0	0.4	−5.7	−7.0	−7.5	−7.75
OECD Europe	−3.3	−23.7	−9.3	−18.0	−9.5	1.0	−21.75	−10.5	−19.25	−9.25
Total OECD	−3.3	−22.75	5.5	−13.0	−5.0	1.0	21.75	−6.5	−22.5	−17.5

[a] Austria, Finland, Norway, Sweden, Switzerland.

[b] Greece, Portugal, Spain, Turkey.

Sources: OECD, *Economic Outlook*, no. 17, Paris, June 1975, Tables 20, 22, 23, pp. 56, 57, 58; no. 18, Paris, December 1975, Tables 25, 27, 28, pp. 61, 62, 63; no. 20, December 1976, Tables 22, 23, 42, 43, pp. 65, 66, 113.

The case of the United States is typical in that the dramatic improvement in the trade balance of 1975 was due in part to the depreciation of the dollar in relation to the currencies of the main trading countries, and in part to the economic recovery of the second half of 1975, which started earlier in North America and in Japan than in Western Europe. The 1976 trade deficit was in part the consequence of the sharp increase in oil imports.

The EEC group of countries registered trade and current-account deficits in 1974, followed by a surplus in 1975 and a situation of equilibrium in 1976. But the global image is misleading. In spite of a very high oil bill, West Germany is accumulating enormous trade surpluses, thanks to the superior quality of its exports of capital goods, but Britain's trade and current balances have been in the red since 1974. At present, Britain imports about as much oil as Germany, though North Sea production will reduce the oil bill in the late 1970s. But Britain's economic performance has been poor, with the highest rate of inflation in the OECD area. France managed to bring its trade balance into the black in 1975, after a substantial deficit in 1974 but was back in the red in 1976. Italy shares with France the handicap of scarce domestic energy resources. After a huge trade deficit in 1974, it was able, because of drastic economic measures, to improve its external position in 1975, but it deteriorated again in 1976. This brief survey shows that the four larger countries of EEC differ in their resources as well as in their policies and performances.

The situation in Northern and in Southern Europe is not as favorable as that of the Common Market as a whole. The smaller countries, with the exception of Switzerland, are registering sizable current balance deficits throughout the 1974–76 period. The setback in tourism and in money brought in by migrant workers will slow down improvement in the Mediterranean countries.

Europe's Trade with OPEC. OECD's exports to OPEC rose from $16.5 billion in 1973 to $28.9 billion in 1974 and to $47 billion in 1975. Corresponding figures for the EEC were $8.2 billion, $13.4 billion, and $22.7 billion. The drive to increase exports by the industrial countries to their suppliers of oil was effective. But the EEC, the United States, and Japan were competing in a relatively narrow market. The EEC countries as a group maintained approximately their share of the OPEC market, 49.7 percent in 1971–73, 46.7 percent in 1974, 48.5 percent in 1975, and 47.6 percent in 1976.[12]

[12] OECD, *Economic Outlook*, no. 17, Table 45, p. 117, no. 19, Table 51, p. 114, and no. 20, Table 46, p. 114.

To assess the immediate impact of the new oil prices on Europe's economy, the balance between the increased cost of oil imports, amounting in 1974 to approximately $30 billion, must be compared with the increased revenue from exports to OPEC, amounting to $6.25 billion in 1974.[13] The residual "deficit increment" deriving from Europe's trade with OPEC amounts to about $24 billion and represents barely 2 percent of the gross domestic product of Western Europe in 1973 ($1,300 billion). Such is the approximate amount of the actual transfer of real income which took place in 1974, a year of practically no growth for Europe's gross national product.

The Trade Balance over the Next Decade. At a time when all industrialized countries are attempting simultaneously to build a non-oil trade surplus, an evaluation of the absorptive capacity of the OPEC countries becomes essential. These countries are using part of their oil receipts to increase imports both of consumer goods, such as food and textiles, and of equipment goods, such as turn-key plants in the chemical, petrochemical, mechanical, and electro-mechanical industries, in an attempt to move from a semifeudal to an industrial age and to set up a modern agro-industrial economy during the expected life span of their oil and gas reserves. The situation varies from country to country according to the relationship between population and volume of proven reserves of fossil fuels. Algeria, Ecuador, Indonesia, Iran, Iraq, Nigeria, and Venezuela have a high absorptive capacity for imports of food and manufactures because of their need to improve the standard of living of a large and fast-growing population (253 million in 1972). By contrast, Saudi Arabia, Kuwait, the United Arab Emirates, and Libya, with an aggregate population of 9.3 million in 1972, have a low absorptive capacity for imports.[14]

OECD has reviewed the medium-term outlook for OPEC's current balance, assuming that the relative price of oil is unchanged from its present level. The organization takes a sober view of the demand for oil from the OECD area, which may be roughly the same in 1980 as it was in 1974. It assumes that the imports of countries with high absorptive capacity may be roughly equal to oil-export revenues by 1980, taking into account the grants and loans that OPEC countries are offering and will be offering to non-oil-producing developing countries.

The conclusions of the OECD survey are that the OPEC current-account surplus may fall from some $67 billion in 1974 to $45 billion

[13] Ibid., Table 21, p. 56.

[14] *United Nations Statistical Yearbook*, 1973, pp. 67-71.

(in 1974 dollars) in 1975, to $25 billion in 1980, and to an approximate balance in 1985. The cumulative surplus may amount to $200–250 billion in 1974 dollars in 1980, and may decline slightly from then to 1985. Finally, the OECD current account may be back in surplus by 1980.[15]

The revised OECD estimates are much lower than forecasts published after the first shock of the 1973–74 price increase and somewhat higher than forecasts published by private banks in 1975. Though most oil-producing countries have shown restraint in the disposal of their surplus income, some of them have been obliged, in view of the decline in demand, to revise downward their overambitious development plans and even to borrow on the financial markets. Such is the case for Iran and Algeria.

What will be the breakdown of the cumulative surplus between the main trading partners of the OPEC countries, the United States, Japan, Western Europe, and the non-oil-producing developing countries? A number of variables must be taken into consideration, such as the national energy policies of the oil importers, the relative par values of the currencies, especially with regard to the dollar, and the possibility that the stronger countries may pass on the burden to the weaker ones. Three elements in particular bear on the situation: the lasting dependence of Western Europe on external supplies, even with a determined policy of energy conservation and energy substitution; the remote possibility that the United States may move toward near self-sufficiency, if it so decides; and the approach of the developing countries toward the limits of their borrowing capacity. In view of these factors it seems likely that a large portion of the accumulated surplus income of the OPEC countries would come to bear on Western Europe.

Payment Mechanisms

Faced with a sudden and heavy current-account deficit, a debtor country, apart from promoting exports and reducing internal consumption, must resort to external borrowing and possibly to the sale of monetary or real assets. Faced with a current-account surplus, a creditor state, apart from expanding imports and domestic consumption, must invest its surplus revenues in either liquid or durable assets. In the situation created by the rise in oil prices, the policies followed by the European countries, individually and collectively, were influenced by the deci-

[15] OECD, *Economic Outlook*, no. 17, p. 78 and Table 33, p. 79.

sions of OPEC countries concerning the disposal of their surplus revenues.

Disposal of OPEC's Surplus Revenues. The monetary authorities of the OPEC countries, especially in the Middle East, were accustomed to investing their surplus funds in the New York and London money markets and in the Eurodollar market. In 1974 and 1975, they pursued this policy of high liquidity preference on a much larger scale. An estimated 19 percent of OPEC's 1974 surplus of approximately $57 billion was invested in the New York market, while 13 percent was invested in the London market, and 40 percent in the Eurodollar market.[16] The United States received a net inflow of petrodollars from the OPEC countries. Britain also benefited, although the inflow of funds was reduced by its economic predicament. Eurobanks absorbed OPEC short-term deposits and re-lent the funds, generally in the form of medium-term credits. As a result, the volume of loans and deposits became disproportionate to the equity resources of the Eurobanks. After the bankruptcy of several smaller institutions, such as the Herstatt Bank in Cologne, lenders tended to concentrate their deposits in the larger banks. Thus the capacity of the market and its role as financial intermediary seemed to have reached its limits.

The oil-producing countries were very selective in engaging their funds on the stock and long-term bond markets. A number of reasons account for their caution: political distrust regarding possible nationalization of foreign assets, the lack of a network of financial and economic information, and the general economic situation in the industrialized countries which combined inflation and recession. Several sizable direct investments were made in 1974, mainly in stronger countries such as West Germany, motivated by the prospects of technological transfers. Kuwait purchased 15 percent of Daimler-Benz. Iran bought 25 percent of Krupp, whose expertise will be used in joint ventures in Third World countries. Iran also took over an indirect 10 percent share in the French-dominated Eurodif consortium which is building a uranium enrichment plant in France. Real estate interests have been bought out in Britain. In the future, portfolio investments may develop, but the purchase of equity will probably be directed more toward the New York Stock Exchange than toward the European financial markets.

To summarize, the OPEC countries have tended to lend to or invest in the strongest of the industrialized countries, and have shown restraint in the use of the Eurodollar market, indicating that

[16] Ibid., Table 27, p. 70.

it was not their responsibility to "recycle" the flow of oil revenues from the most-favored to the least-favored consumer countries.

Policies of the European Countries. These countries were reluctant to tap their monetary reserves of gold at a time when the official price of gold was far below the market price. They preferred to resort to foreign borrowing on a national basis. The member countries of EEC also practiced a degree of mutual aid.

The situation and policy in regard to external financing varied from country to country and time to time. France borrowed on the Eurodollar market in 1974, registered a surplus in its foreign payments in 1975, and was again an external borrower in 1976. The position of Britain deteriorated in 1974 and again but to a lesser extent in 1975. A line of credit of $2.5 billion in Euromoneys opened in 1974 was fully utilized by mid-1975. Italy increased its foreign indebtedness in 1974 by a total of $5.5 billion lent by the International Monetary Fund (IMF), by its EEC partners, and by West Germany. Its external position further deteriorated in 1975, necessitating recourse to the IMF in the form of the "Oil Facility" (see below).[17] Meanwhile, West Germany remained throughout the 1974–76 period a strong creditor country.

For many years, the member states of EEC failed to implement the goal they publicly stated in 1969 of setting up an economic and monetary union. The reasons were twofold: the disarray of the international monetary system since 1971, and the diverging national economic and monetary policies. In April 1973, EEC at last succeeded in setting up the European Fund for Monetary Cooperation, which helped Italy face its balance-of-payments difficulties in 1974 by granting a short-term loan, later converted into a medium-term arrangement. The EEC Council of Ministers also decided in October 1974 that the European Community as such was entitled to collect resources abroad from oil producers for amounts up to $3 billion, in return for granting medium-term loans to its weaker member states. This plan has not yet been implemented.

International Banking and Monetary Cooperation. These EEC measures, however, were obviously insufficient in amount and scope to deal with the recycling problems of the industrialized countries, not to mention those of the non-oil-producing developing countries.

[17] *Bank for International Settlements, 45th and 46th Annual Reports* (Basel, 1975), pp. 88, 91, 93 (English edition); 1976, pp. 72, 74 (French edition).

Some initiative in the monetary field was taken by the Group of Ten, composed of the monetary authorities of the ten wealthiest members of OECD. In September 1974, the central banks of the Group of Ten decided to act jointly as lenders of last resort to the Eurobanks and to tighten their control over Eurobank operations. But the private financial institutions had to be supplemented by recycling facilities under the aegis of international organizations, and therefore under government sponsorship. To this end, the Group of Ten in June 1974 established the "Oil Facility" of the International Monetary Fund, a clearinghouse for oil surpluses and deficits. The IMF offered producer countries a financial asset in the form of Special Drawing Rights (SDRs), and lent consumer countries up to approximately 3 billion SDRs in 1974 and up to 5 billion SDRs in 1975 (the equivalent of approximately $3.5 billion and $6 billion).

A further initiative came from the United States. Aware of the limits of existing recycling procedures, Secretary of State Henry Kissinger proposed to set up a Financial Support Fund for OECD member states in the amount of $25 billion. The fund would act as a safety net "in exceptional circumstances," with support from the fund contingent on reasonable self-help measures by the borrowing countries. Each participant would be allotted a quota as a basis for its financial contribution, its borrowing capacity, and its voting rights. The agreement creating the fund was signed in April 1975, and is subject to ratification by the parliaments of the OECD member states.

Up to the present, the needs of the non-oil-producing developing countries have not been met. A special trust fund in the IMF for the benefit of the poorest countries was not scheduled to become operational until the 1976 annual meeting of the fund. Several OPEC countries have set up national and multinational specialized funds with a view toward helping preferably Arab or Moslem countries. In 1975 the official development assistance benefiting non-oil-producing developing countries amounted to $14.75 billion, of which $11.5 billion was granted by OECD countries and $2.5 billion by OPEC countries. The greater part of this aid is accounted for in the current-account deficit of this group of countries, estimated at $26.5 billion.[18]

On the whole, the international banking system has shown a remarkable resilience in dealing with large sums of petrodollars, thanks to the predominance of the industrial countries. Their commercial banks control the Eurodollar market, and the governors of their central banks, with a long experience of cooperation, meet in Basel at the seat of the Bank for International Settlements. This

---

[18] OECD, *Economic Outlook*, no. 19, Table 52, p. 115.

contribution of the industrial countries has been aided by the caution and restraint of OPEC monetary authorities, whose role in the IMF will develop with the impending increase in the quotas of their respective countries.

Although the immediate recycling problems seem to have been solved without any major accident, no valid solution has yet been developed for financing current-account deficits of the non-oil-producing developing countries in the coming years.

Trends in Prices

Oil prices are governed by the combination of an oligopolistic market structure and political forces in a geostrategic setting. It is with this dual approach in mind that one should analyze the impact of inflation in the short term and the strength of OPEC as a price-fixing cartel in the medium term.

Inflation, Recession and the Short Term. Consumers complain of inflation induced by higher oil prices, and producers complain of the reduction in the real price of oil because of inflation in industrialized countries. Both complaints are relevant.

OECD evaluated the direct impact of oil price increases at 1.5 percent in the total domestic expenditure deflator of OECD countries in 1974, and at 2.3 percent for Europe alone.[19] This compares with a GNP/GDP deflator in 1974 of 12.2 percent for total OECD and of 11 percent for OECD Europe.[20] Roughly speaking, the rise in oil prices may have contributed one-fifth of the present inflation in Europe.

Apart from the depreciation in the exchange value of the dollar, the currency in which oil prices are expressed, the erosion of purchasing power of OPEC's revenues derives from the rise in the price of goods and services that the oil producers import from OECD countries. This rise is greater for imports from Europe than for imports from the United States. Since oil prices reached $9.31 per barrel in January 1974, OPEC's real income has fallen by roughly a quarter.[21] World recession led to a reduction in the volume of oil imports in the OECD area of 3.5 percent in 1974 and 8 percent in 1975 and to a timid return to a buyers' market.[22] (Corresponding

[19] OECD, *Economic Outlook*, no. 15, p. 33.

[20] OECD, *Economic Outlook*, no. 17, Table 37, p. 114.

[21] "Two Years of OPEC," *The Economist*, September 20, 1975, p. 83.

[22] OECD, *Economic Outlook*, no. 18, Table 22, p. 57; no. 19, Table 20, p. 48.

figures for Western Europe were 7.6 percent in 1974 and 11.4 percent in 1975.)[23] A 10 percent price rise was decided on in September 1975, and there was a question whether the market would bear it. However, at the end of 1975 economic recovery began in the United States and gradually got under way in Europe and Japan. As a result, oil consumption slowly rose in 1976, while the general level of crude oil prices remained unchanged until December 1976, when the price of oil was raised 10 percent by all OPEC countries except Saudi Arabia, which refused to increase its selling price by more than 5 percent.

As the OPEC countries have a surplus production capacity on the order of 10 million barrels per day, their revenues will probably benefit more from the normal recovery of demand than from an excessive price hike. However, to quote Colin Robinson of the Henley Centre for Forecasting, "It is no exaggeration to say that the biggest contribution to lower prices which the governments of the industrial countries could make would be to control their own rate of domestically generated inflation."[24]

OPEC as a Price-Fixing Cartel in the Medium Term. OPEC is stronger than any producer cartel in the past for a variety of reasons.

The first is the position of oil as the most important source of energy in use in the world and the most important commodity dealt with in international trade. In 1973 OPEC countries controlled 90 percent of world crude-oil exports and 71 percent of proven world reserves.[25]

The second reason is that the price elasticity of demand is low, at least in the short term. Petroleum products are nonsubstitutable fuels for road transportation at a time when automobile use and ownership have become a way of life in the industrialized countries. As mentioned above, the percentage decline in demand in 1974 and 1975 was minimal compared with the 1973–74 quadrupling in prices.

Third, the short-term price elasticity of alternative supplies either of non-OPEC oil or of substitute primary sources of energy is low because of the lead time in their production processes.

Finally, one country with a dominant position in OPEC can ensure the cohesion of the group at least for some time. OPEC member states have not signed any formal agreement on production control, but Saudi Arabia was able, by curtailing its own production, to respond to the decline in demand that has taken place since 1973.

[23] "Slump in Crude Oil Imports," *The Petroleum Economist*, June 1976, p. 215.

[24] Robinson, *Energy Depletion*, p. 28.

[25] *BP Statistical Review*, 1973, pp. 4, 10.

OPEC countries were selling 30 million barrels per day in 1973. Production had fallen back to 26 million barrels per day in January–February 1974. Saudi Arabia, with a 12 million barrel-per-day capacity, cut its production from over 8.4 million barrels in 1974 to 6 million in April 1975 and could cut it further.[26] The implementation of its long-range development plan for a small population is not dependent upon the maximization of crude-oil production. The strength of OPEC is therefore linked to the flexibility of Saudi Arabia's production policy, which is able to cushion the variations in demand from the oil-consuming countries. If imports from OPEC were reduced to a level of 20 million barrels per day, then most of the members of OPEC would have to resort to production control, and the cartel might break up. (Another cause for disbandment would be a political upheaval in Saudi Arabia, a matter to be discussed in Chapter 5.)

The basic question is whether and when the oil-consuming countries will reach the goal of reduced imports. On the supply side, much will depend on the speed of the substitution programs of the OECD countries, in other words, the contributions of non-OPEC oil and gas and of nuclear energy. An assessment of Western Europe's potential in this respect is made in Chapter 4. On the demand side, two factors will intervene: the level of economic activity in the industrialized countries, and the seriousness of the energy conservation programs they would be well advised to launch.

Taking into account all these variables, it seems that in the medium term the price elasticity of demand for oil and the price elasticity of supply for non-OPEC oil and for substitutes will increase. Then OPEC's bargaining power may be eroded—but only to a point, because the oil-consuming countries must protect their sources of high-cost alternative energy.[27] If oil-consuming countries allow oil imports to fall below a certain price, then the profitability of past investments in substitutes would be impaired, and the capital required for further investments could not be raised. Hence the United States proposed the establishment of a floor price for oil submitted to the members of the International Energy Agency (IEA) in Paris. There were conflicting views on this proposal between potentially oil-rich countries, such as Britain and Norway, who were eager to protect their North Sea investments, and the oil-deprived countries, such as Italy and Japan. Nevertheless, the IEA at its February 1976 board

[26] "The Gulf, a Survey," *The Economist*, May 17, 1975, pp. 15-19; "Two Years of OPEC," *The Economist*, September 20, 1975, p. 84.

[27] Ian Smart, "Uniqueness and Generality," in *Daedalus*, Fall 1975, p. 266.

meeting adopted a recommendation in favor of a floor price of $7.00 a barrel.[28]

In short, OPEC is a strong cartel indeed, and the counter strategy of the consumer countries can be effective in the medium term only within the limits set by the floor price. In the long term, energy prices are bound to rise in relation to the exhaustion of the low-cost deposits and to the corresponding increase in capital output ratio.

Conclusion

Higher energy prices cause a double transfer of resources—from consumption to investment inside the industrialized countries and externally from the industrialized to the oil-producing countries.

The inside transfer reduces the real disposable income of the wage earner and requires shifts of manpower and capital from one industrial branch to another. With a period of relative scarcity succeeding one of affluence, will the European wage earner accept a constriction in his life style, or will he fight for a higher nominal income? The answer so far has been resistance rather than acceptance, except in West Germany. Sacrifices are not welcome in consumer societies, where public spirit is at a low ebb, nor are they easily enforceable by weak governments.

In the field of international relations, the Common Market countries, with the major exception of Britain, were able to restore the equilibrium of their current balance in 1975. The recovery was short-lived for France and Italy whose situation deteriorated again in 1976. The smaller countries of Northern Europe did not fare so well, and those of Southern Europe even less so. But the most serious situation is that of the non-oil-producing developing countries, which are incurring unbearable current balance deficits. These countries will sooner or later have to curtail their imports of manufactured goods from the industrialized countries, and notably from Europe. At the same time, with the help of Western technology, oil-producing countries are building some basic industries that will compete directly with their European opposite numbers. The rise in energy prices is therefore the signal both of a contraction in world trade and of a new international division of labor. Europe will have to revise its whole industrial strategy.

The recycling of petrodollars has been manageable so far, thanks to the proper functioning of the international banking system and to

---

28 "IEA Agrees Floor Price for Imported Oil," *The Petroleum Economist*, March 1976, p. 104.

the restraint of OPEC monetary authorities. But in the absence of a European economic and monetary union, it is doubtful that West Germany would be in a position to back the weaker countries of Europe if their financial situation further deteriorated. The dollar remains the dominant currency, and the safety net proposed by the United States has no European counterpart. Efficient energy substitution and conservative policies, with the view of limiting the price-fixing capacity of OPEC, require strong action by all industrial countries.

In sum, economic policy problems related to energy cannot be solved in a geographical setting limited to the relations between the European countries and their main suppliers of oil in the Middle East and Africa. These problems also involve the non-oil-producing developing countries and the United States. They encompass major political issues in a worldwide strategy.

3
NATIONAL POLICIES

Policy Objectives and Energy Balance Sheets

The general complacency in management of energy resources for Western Europe as a whole, analyzed in Chapter 1, is a reflection of the policies of each of these countries. Any national energy policy must be influenced by the characteristics of the different energy sources and by the relationship between the nation's consumption and its supplies, both domestic and external.

In a competitive market, the economic optimum corresponds to the equalization of marginal costs of competing energy sources—regardless of the origins of the sources—whether the equalization of costs is realized by the market mechanism or by state intervention. Such is not the case in an oligopolistic industrial structure. Furthermore, the search for lower costs may not be compatible with the political optimum of securing supplies over the long term. Two questions arise that involve difficult policy decisions: (a) What ratio of external supplies to total domestic consumption will ensure a reasonable amount of security? and (b) What cost, in terms of higher domestic energy prices, is any given country prepared to pay to approach the optimal ratio?

As a matter of fact, most European energy policy makers failed to perceive the relationship between the economic and the political optimum, between higher prices and higher political-economic security. Since low-cost energy contributed to overall economic growth, it was given priority over other considerations. The political implications of highly concentrated energy resources in the Middle East were not understood in Europe until the steep rise in prices and the temporary reduction in deliveries showed that oil was a major weapon in power politics—and one not in the hands of the Europeans.

The rise in energy prices introduced new constraints on national energy policies in terms of higher investments in energy conservation and substitution, redeployment of industry, the return of balance of payments to equilibrium, and foreign trade orientation, as explained in Chapter 2. The increased number and wider range of energy policy objectives imply difficult trade-offs and encompass the overall economic strategy of the state.

It is within this framework that the national energy balance sheets of the European countries must be analyzed. Table 7 presents highlights of the economics of energy in ten selected countries for the year 1973, a peak consumption year, which came at the outset of the oil crisis. These countries should be studied and compared in terms of the management and availability of resources.

France has had a long tradition of state intervention in the oil industry. Britain and the Netherlands, the seats of two powerful multinational oil companies, relied on the operations of the private cartel for their supplies. A late-comer to the field, Italy built its oil and gas industries from a state-owned company. Coal and gas were put under state control in France and in Britain. These two countries and others devoted large public appropriations to research and development in nuclear-generated electricity, and came to supervise the various branches of the nuclear industry. West Germany stuck as long as it could to the free market system, but was obliged to depart from it in the aftermath of the coal crisis and at the height of the oil crisis. Environmental considerations reinforced state control in all countries.

Thus energy industries in Europe are not only oligopolistic, as explained in Chapter 1, but also more and more subject to state intervention. Therefore, they are highly compartmentalized in their medium-sized or small national markets.

Turning now to the question of availability of resources, it appears that the economies of West Germany and Britain are still meeting one-third of their total primary energy consumption with domestic coal and are dependent on imports for one-half of their total consumption. The two other large industrial countries, France and Italy, have small domestic energy resources, and their economic development is based on massive oil imports. The country with the highest level of self-sufficiency is the Netherlands, because of its large natural-gas reserves. Britain and Norway will reach complete

Table 7

FEATURES OF ENERGY ECONOMIES OF EUROPEAN COUNTRIES, 1973

(million tons of oil equivalent)

	West Germany	France	Italy	Netherlands	Belgium	U.K.	Ireland	Denmark	Sweden	Norway
Production	120.08	42.49	26.61	57.82	6.41	112.25	1.24	0.07	17.47	12.83
Net imports [a]	146.65	139.76	105.49	4.01	40.83	111.55	5.89	19.95	30.10	6.71
Total energy requirements	266.73	182.25	132.10	61.83	47.24	223.80	7.13	20.02	47.57	19.54
Dependence on foreign supply (%)	54.98	76.68	79.85	6.48	86.43	49.84	82.60	99.65	62.27	34.33
Share of different types of primary energy (%)										
solid fuel	31.06	17.13	7.78	4.69	25.31	35.24	22.15	12.54	10.15	3.78
crude oil & petroleum products	55.91	67.22	73.54	48.08	58.93	49.55	75.31	87.51	59.78	42.53
natural and industrial gas	10.29	7.56	10.97	46.96	15.49	11.45	—	—	—	—
primary electrical energy [b]	2.72	8.07	7.70	0.22	0.25	3.75	2.38	—	30.03	53.63
Energy consumption per capita	4.30	3.40	2.30	4.00	4.70	3.90	2.40	3.90	5.20	4.00

[a] Including marine bunkers and change in stock.

[b] Nuclear, hydro and geothermic power plus electricity net imports.

Source: OECD, *Energy Balances of OECD Countries 1960-1974*, Paris 1976 and *UN Statistical Yearbook 1974.*

self-sufficiency within a few years, thanks to the development of North Sea oil and gas.

Denmark and Ireland have practically no domestic resources. Those of Belgium are scarce, those of Sweden somewhat greater, because of hydroelectricity. Among the countries not listed in Table 7, Spain and Portugal are in a position of external dependence comparable to that of France and of Italy. Turkey can expect to improve its position through the exploitation of its sizable coal reserves. Greece has good prospects of flows of offshore oil in the Aegean Sea.

Since North Sea hydrocarbons and nuclear-generated electricity are expected to become important factors in the improvement of the energy balance sheet of European countries over the next decade, the following study of national policies will deal with (1) the countries endowed with large oil and gas reserves, that is, Britain, Norway, and the Netherlands, and (2) the countries that are banking on the development of a nuclear industry, that is, West Germany, France, Italy, and Sweden.

Great Britain

Britain's industrial supremacy in the nineteenth and twentieth centuries was linked to the availability of cheap coal. Britain was the first European country to foresee the importance of oil and to invest in the Middle East through the Royal Dutch Shell group and through the Anglo-Iranian Company, later given the name of British Petroleum (BP). Substitution of oil for coal took place in the 1960–73 period, when the share of coal fell from 73 percent of total energy consumption to 35 percent, and that of oil rose from 26 percent to 49 percent.[1] The substitution process may have been accelerated by favorable prospects for North Sea oil and natural gas. These prospects lessened the urgency for development of nuclear power, initiated some twenty-five years ago.

The role of the state in the British energy sector is overwhelming. Coal is nationalized. So are the distribution of industrial and natural gas, under the British Gas Corporation, and the production and distribution of electricity under the aegis of the Central Electricity Generating Board (CEGB). The Atomic Energy Authority (AEA) makes the major decisions in the nuclear field. The state owns 49 percent of the capital stock of BP but takes no direct part in its management. The 14 percent ownership of Burmah Oil included in the capital stock

[1] OECD, *Energy Balances of OECD Countries 1960-1974* (Paris, 1976), pp. 431, 444.

of BP has been taken over temporarily by the Bank of England, in order to cope with the financial difficulties encountered by Burmah Oil in 1974. In addition, as explained below, the British government has been increasing its grip on concessions in the North Sea.

Britain's policy has been geared to retarding the retreat of coal by a number of protective devices: subsidies to the National Coal Board, prohibition of coal imports, and taxation of petroleum products used by electrical power stations, industry, and households. But successive large increases in the wages of miners have hampered the profitability of coal mines, even after the steep rise in oil prices. Coal production has fallen from 196 million tons in 1960 to 114 million tons in 1973. Economically recoverable reserves amount to 3.871 megatons.[2] The coal industry envisages the opening by 1985 of new mines in the Midlands and in Yorkshire with a total capacity of 40 million tons per year, to replace worked-out ones.[3]

Britain is at present the second largest producer, after the United States, of electricity from nuclear sources, but it will soon be overtaken by the Soviet Union, Japan, West Germany, and France. The first generation of reactors, the Magnox, was graphite-moderated, gas-cooled, and fueled by natural uranium. It was abandoned in the 1960s because it was relatively inflexible and expensive. Nine stations were built, with a combined generating capacity of 3,700 megawatts (MWe). The successor to the Magnox was the Advanced Gas Cooled Reactor (AGR) using enriched uranium. A program of production based on the building of five power stations encountered a number of technical difficulties, before being rescheduled to come into operation by 1978, adding some 6,600 MWe of capacity.

The British government rejected a proposal of the Central Electricity Generating Board for the installation of new generating capacity based on the American Light Water Reactor (LWR), selected by a number of European countries either in the pressurized-water version (PWR) or in the boiling-water version (BWR). Instead, in 1974, the Steam Generating Heavy Water Reactor (SGHWR), colloquially called the steamer, was chosen, employing enriched uranium, with the required heavy water as a moderator to be supplied by Canada. This government choice came as a surprise. Was it the result of anxiety about hazards that had led to the temporary abandonment of some of the light water reactors in the United States? Or did the British government doubt the technological capacity of its industry to catch

2 World Energy Conference 1974, *Survey of Energy Resources* (N.Y.: The U.S. National Committee of the W.E.C., 1974), p. 315.

3 OECD, *Working Group on the Reappraisal of the Role of Coal*, 1975, p. 9.

up with experienced U.S. and European competitors in the export markets? Was the decision a measure to protect British industry by adopting a specific type of reactor? Whatever the case may be, since only one 100 MWe experimental SGHWR is in operation today, the development of an important program will take a number of years. A 4,000 MWe generating capacity is to be initiated over the next four years.[4]

In 1959 the Atomic Energy Authority installed an experimental fast breeder reactor in Dounreay, now linked to the Scottish electricity network. A plan to order a 1,300 MWe breeder in 1976, to become operational around 1983, is at present under consideration.

The British nuclear program has developed over twenty years, using domestic technical resources. In spite of successive reorganizations, it has failed to create a strong industrial structure capable of producing commercial returns abroad. Several industrial teams that had been formed to design nuclear power stations were merged into two groups in 1969, then in 1974 into one single unit, the National Nuclear Corporation, in which General Electric (U.K.) and the Atomic Energy Authority, respectively, hold 50 percent and 15 percent of the capital stock.[5]

Britain has no uranium ore in its territory. It imports most of its enriched uranium from the United States and the remainder from the Soviet Union, while engaging in enrichment centrifuge technology jointly with West Germany and the Netherlands. If difficulties develop with centrifuge technology, the CEGB will apply for a larger supply from the Soviet Union. Meanwhile, Britain is associated with West Germany and France in the reprocessing of nuclear fuels.

Exploration of the British sector of the North Sea began in the early 1960s, resulting in major gas discoveries in the southern area in the mid-1960s. The first major oil field was confirmed in 1971 in the northern area, east of Scotland. In 1976 official estimates amounted to 1,350 million tons of oil for proven reserves from licensed areas, and to 4.5 billion tons for total recoverable reserves from presently designated areas, not all of which are as yet licensed. Corresponding figures for gas were 29 and 51 trillion cubic feet (700–1,100 million t.o.e.). Current discoveries could support production during the 1980s of 100–150 million tons of oil and of 5,000–6,000 million cubic feet per day (mcfd) of gas (44–53 million t.o.e.).

---

[4] "Britain's Nuclear Choice," *The Petroleum Economist*, December 1974, p. 449.

[5] Assembly of Western European Union, *A European Policy on the Peaceful Uses of Nuclear Energy*, nos. 36, 37, (1974), p. 8.

Production of gas from the southern basin in 1974 averaged 3,500 mcfd and provided about 95 percent of British consumption. The first oil output from the northern basin was delivered in 1975, and the pipeline from the Forties field to Scotland was completed in the same year. It is expected that by 1980 oil production will amount to 95–115 million tons, enough to make Britain self-sufficient.[6]

The British government's declared policy is to ensure that oil production builds up as quickly as possible to and above the self-sufficiency level. Some oil would be available for exports from 1980 onward in the form of light crude oil to be sold to Continental Europe against the purchase of heavy crude oil from the Middle East. The volume of net exports will depend both on the rate of consumption and on the volume of new discoveries. Without major new discoveries, Britain will again become a net importer by the 1990s. The impact of North Sea oil and gas on the balance of payments in the next decade is a complex matter. On its petroleum account, Britain will save increasingly in net imports and later benefit from net exports. On its non-oil account, imports of equipment for the development program are probably at their peak, but financial outflows such as interest payments, remittance of profits, and repayment of principal will increase over the years, since 60 percent of the assets of the commercial fields are in the hands of foreign companies. According to recent estimates, the potential effects of the North Sea oil and gas programs on the balance of current payments would amount to £9,500 million in 1980 and to £24,000 million in 1985. These figures are based on the assumptions that the 1976 price of oil will be maintained in real terms and that the North Sea resources will be used for balance-of-payments improvement as against domestic absorption such as extra investment. In such case the net effect on current account would represent 5.4 percent of the gross national product in 1980 and 7.9 percent in 1985.[7]

The capital cost of field development has escalated under a variety of factors, such as delays, supply shortages, and cost inflation. The average cost per barrel-a-day capacity for North Sea oil was £2,500 as against £100 in the Middle East. If Britain's own inflation remains above the average of the industrial countries, the margin between development costs and the price of oil may be eroded.[8]

The original licenses for North Sea development were granted on very favorable terms. After a swing in the other direction, a reason-

[6] U.K. Department of Energy, *Development of the Oil and Gas Resources* (London), 1975 and 1976 reports.

[7] *Economic Progress Report*, Information Division of the Treasury, London, no. 76, July 1976, Tables 1 and 3, pp. 1 and 3.

[8] Ibid., p. 16.

able compromise has been reached, with taxes consisting of a royalty of 12.5 percent of the wellhead value of the oil, a petroleum revenue tax (PRT) at a rate of 45 percent, and the regular corporation tax on net revenue. Tax regulations will be revised in case of a fall in oil prices.

In 1975 the British National Oil Corporation (BNOC) was established, to produce, refine, and market North Sea oil and to control state participation in offshore fields, which from this point on amount to 51 percent of the total operation. The implementation of the government corporation required renegotiation of existing agreements, now well under way. A far-reaching agreement for participation of BNOC in BP activities took place in 1976 with BNOC scheduled to take part in upstream as well as downstream activities.

British industry, however, failed to provide the new technology for offshore oil developments. None of the pipelines has been supplied by Britain. There are no British designs for drilling rigs. Production platforms have been supplied principally by U.S. and European companies, while most of those built in Britain are of foreign design. According to a recent survey, only 33 percent of offshore expenditure has accrued to British companies.[9]

The majority of British oil fields are on the continental shelf of Scotland. As a result, the North Sea coast of Scotland, with Aberdeen as a capital, provides the best location for oil-related industries and supply bases. Unfortunately, the resulting increase in employment in Eastern Scotland barely compensates for the corresponding decline in the west, especially in the Clydeside area, where the shipbuilding industry has lagged in initiating new technology for oil. The North Sea developments have fueled the claims of the Scottish National party, which in 1974 became the second largest party in Scotland. The British government is committed to the creation of a Scottish assembly and is considering "devolution," that is, a still undefined degree of regional autonomy. But an independent Scotland overflowing with oil revenues would have a considerable balance-of-payments surplus and could pursue an expansionary economic policy.[10] Its position would be comparable to that of Norway. Thus a major British domestic political problem has arisen from the oil discoveries.

In the international field, the British government strives to keep maximum liberty of action in terms of production volume and of prices for North Sea oil. But balance-of-payments constraints will

---

[9] Ibid., pp. 21-24.

[10] D. I. MacKay and G. A. MacKay, *The Political Economy of North Sea Oil* (London: Martin Robertson, 1975), p. 171.

probably induce the government to extract oil at the fastest possible rate, contrary to the fears of Britain's partners in the EEC. Similarly, it is doubtful whether Britain will endeavor to establish a domestic price for oil lower than that for its surplus on the Continent, a discriminatory measure prohibited by the Treaty of Rome.[11]

In the realm of investment, the British have asked the Community to contribute to the financing of marginal fields in the North Sea. In response, the Community agreed to help finance a number of projects in 1975 and 1976, but only for very small amounts of money ($40 million in 1975). These agreements, in spite of their limitations, should pave the way for the establishment of a sharing system for the surplus of light crude oil which Britain is expected to produce in the coming years.

At the same time, Britain did not hesitate to join the International Energy Agency in 1974. Within IEA, Britain has strongly supported a U.S. proposal for a floor price of oil, aimed at protecting investments in the North Sea. In general, Britain will carry more weight in both EEC and IEA in relation to the development of its domestic oil and gas resources.

Britain has not always succeeded in avoiding conflict with its EEC partners. During preparations for the North-South conference on energy and raw materials, which was to be held in Paris in December 1975, Britain refused to be represented by the European Community and claimed an individual seat. Eventually the claim had to be dropped.

Britain has taken a pro-Arab stand in the Middle East conflict since 1967, when the government sponsored Resolution 242 of the United Nations, advocating the withdrawal of Israel from occupied territories. After the October 1973 war, Britain was declared a friendly country by members of the Organization of Arab Petroleum Exporting Countries (OAPEC). On the specific matter of the Euro-Arab dialogue, a French-sponsored initiative involving the EEC and the Arab League, Britain adopted a cautious attitude in consideration of U.S. policy. Britain's generally favorable attitude toward the Arab cause is linked to its oil and industrial interests in the Middle East, and to the role of the London money market in conjunction with oil. Before and after the 1973–74 crisis, a large portion of the surplus funds of the oil-producing countries was directed either to the money market or to the Eurocurrency market in London. Britain's financial

[11] William Wallace, "Europe: the Changing International Context, Implication for British Policy," *World Today*, May 1975, p. 194.

difficulties have induced several oil-producing countries of the Gulf to discontinue their short-term investments in London.

Norway

Prior to the North Sea discoveries, Norway was a country well endowed with renewable energy resources. Hydroelectricity accounted in 1973 for 54 percent of total domestic energy consumption.

Oil and gas recoverable reserves have almost doubled since the discovery of a giant oilfield, Statfjord, in 1974. They are estimated at approximately 1 billion tons of oil and 700–800 billion cubic meters of natural gas, or 600–700 million t.o.e.[12] Potential reserves are probably much higher since the allocation of blocks for drilling rights has been less rapid in the Norwegian than in the British sector, and no exploration permits have been delivered north of 62° latitude, an area where state sector boundaries are not determined in detail, and where sediments to a thickness of several thousand meters have been detected.

Norway, with a population of only 4 million, is a highly developed industrialized country with a rapid growth rate and full employment. Expansion of the petroleum and natural gas industries will require resources in manpower and—for a few years at least—in capital, which are already employed in other sectors of the economy, especially in industries exposed to international competition. The government fears that too rapid growth of the oil industry would undermine traditional activities in shipping, shipbuilding, fishing, and engineering. Furthermore, the government is not prepared to authorize large-scale immigration to meet the demands for labor of the oil industry, except for workers from the Scandinavian countries, which have established a free-labor market. Under the circumstances, the government is prepared to authorize only a moderate scale of production of oil and gas, and to introduce only slowly the use of revenues from such production into domestic consumption. All phases of petroleum operations are to be under public control.

Production from present reservoirs in Norway may reach 100 million tons of oil and 100 million t.o.e. of gas by 1990. But the government and parliament have agreed to put a ceiling on exploitation. The production goal was set at 50 million tons of oil and 40 million t.o.e. of gas by 1980, about ten times the present domestic consumption of

[12] M. Kvinnsland, "Economic Aspects of Norwegian Petroleum," *Revue de l'énergie*, November-December 1975, p. 126.

hydrocarbons.[13] This figure has been further reduced, and a peak production of hydrocarbons of 75 million t.o.e. is now foreseen by about 1985.[14] At that rate of production, proven and probable reserves may last a century or more. Nonetheless, even this limited production will give Norway a sizable balance-of-payments surplus for the rest of the twentieth century. The country was a net exporter of oil in 1975, and the gross national product is expected to grow by 6 percent per year until the early 1980s.

As Norway is determined not to introduce a high volume of revenue into domestic finances, the most likely use of its surpluses is investment abroad, either direct or portfolio. In the more distant future, money invested abroad will return in the form of interest and dividends, increasing either monetary reserves or domestic consumption. Accumulation of oil income will help the government reduce taxes and expand public expenditure in social services, such as health, education, and housing.

Norsk-Hydro, a conglomerate with its principal activities in the chemical industry, was among the first domestic companies to engage in North Sea exploration. Although the state owns a 51 percent share in Norsk-Hydro's equity, the Norwegian government decided in 1972 to create a new fully state-owned company, Statoil, entitled to hold a majority share in all North Sea concessions. Statoil has the potential of becoming a fully fledged, integrated oil company. It has a 50 percent participation in Statfjord and also in Norpipe, the Ekofisk pipeline to Britain. It will act in cooperation with foreign entrepreneurs who will have a preferential right to oil. For strategic reasons, Statoil will itself undertake all operations north of 62° latitude, in the proximity of the Soviet sector of the Barents Sea, but it will nonetheless propose participation agreements with foreign oil companies. Since an underwater trench parallel to the coast line in the Norwegian sector prevents a pipeline link between the major oil and gas fields and the Norwegian mainland, the Ekofisk gas will be sent to Emden (Germany) and the oil to Teeside (Britain). The two pipelines are now completed. Norway has also agreed to help finance a gas pipeline from Frigg to Scotland.

Norwegian tax provisions on oil and gas concessions, revised in 1975, comprise royalties (from 8 to 16 percent), a corporation tax (50.8 percent), and a special 25 percent windfall profit tax, with the

[13] Bjorn Skogstad Aamo, "Norwegian Oil Policy, Basic Objectives," in *The Political Implications of North Sea Oil and Gas*, Martin Saeter and Ian Smart, eds. (Oslo: Universitetsforlaget, 1975), p. 87.

[14] "Norway, Lower Production Estimates," *The Petroleum Economist*, May 1976, p. 186.

total averaging 57–66 percent. Concessionary companies are not permitted to offset losses incurred elsewhere in the world against Norwegian tax liability. These tax provisions are considered to strike a reasonable balance between state benefit and private profit.[15]

Norwegian private enterprise has moved rapidly in the development of the offshore oil industry. Aker has become a successful competitor to the Americans in rig building, while Norwegian shipbuilders are building a fleet of 150 supply ships.

The Statfjord discovery may lead the government to revise upwards its present production target. Some parliamentarians are now urging the exploration of the basin north of the 62nd parallel, to diversify the economy of the northern provinces.[16] Such a move would be welcome to Norway's chief trading partner, the European Community, which has been critical of the restrictive oil policy practiced so far. The Community will still be largely dependent upon imports from OPEC member states in 1985, but will put pressure on Norway through the International Energy Agency, in which Norway holds the somewhat ambiguous position of an associate member.

As a member of NATO, Norway will need moral support from its allies if it is to exploit its sector of the continental shelf of the Barents Sea. This area is of strategic importance to the Soviet Union, since its civilian and naval traffic from and to Murmansk and Arkhangelsk has to pass through Norwegian waters. Another focus of tension between Norway and the Soviet Union is the Spitsbergen Islands, which are under Norwegian sovereignty, but where the Soviets are exploiting coal mines, in disregard of Norwegian regulations. The Soviet Union and Norway are at odds over the boundaries of their respective sectors of the Barents Sea. They disagree on the interpretation of the 1920 treaty concerning the status of the Spitsbergen archipelago, the Norwegians contending that it is part of Norway's continental shelf.[17] Delicate negotiations between Norway and the Soviet Union were begun in November 1975, while Statoil started drilling north of the 62nd parallel in 1976 to assert Norway's rights.

The Netherlands

The Netherlands has an economy geared towards international trade. Dutch capital holds a 60 percent share in the Royal Dutch Shell

[15] "A Survey of North Sea Oil," p. 33.

[16] Ibid., pp. 34, 37.

[17] "Norvége—des négociations avec L'URSS sur le partage de la mer de Barentz reprendront au mois de novembre," *Le Monde*, September 20, 1975, p. 5.

group, which gave the country direct access to supplies of crude oil overseas, in the Middle East and elsewhere. In 1960 the freely competitive Dutch energy market was divided equally between imported oil and domestic coal. Here as in other coal-producing European countries, oil by 1965 had taken over more than two-thirds of the market at the expense of coal, which retained only 28 percent.

From that point on, the Dutch energy market underwent a spectacular upheaval with the introduction of natural gas from the Groningen field. By 1973 the share of natural gas had become equal to that of oil, that is, 47 percent. The share of coal had fallen to 4 percent, and by the end of 1973, the mines had been closed down. In 1974, for the first time, the Netherlands became a net exporter of energy.

The discovery of the Groningen field was made in a fifty-fifty joint venture between two major oil companies, Esso and Royal Dutch Shell. Production has been entrusted to a specialized subsidiary 40 percent owned by the Dutch State Mines. Transport and sales of gas in the Netherlands and decisions on export policy are under the responsibility of a utility company, Gasunie, the capital stock of which is divided among several entities: the two oil companies (50 percent), the Dutch State Mines (40 percent), and the state itself (10 percent). Thus the private and public sectors are closely associated from the well to the industrial consumers and public distributors. The ultimate authority in pricing policies lies with the Dutch government, which controls transfer prices and approves tariffs. To introduce gas smoothly, without dislocating the energy market, the initial price of gas was set slightly lower than that of oil. Since 1972 it has been linked to the price of oil with low sulfur content. By 1975 the Dutch had achieved their goal of equating gas to oil on the domestic market.

Owing to the extent of the Groningen gas reserves and to the availability of oil throughout Europe in the late 1960s, the Dutch government decided to allocate part of its production to exports. Contracts were signed with national and regional gas-distribution monopolies in France, Belgium, and West Germany beginning in 1967 and in Italy beginning in 1974. In 1973 exports accounted for 48 percent of total domestic production. In December 1974, however, the Dutch informed their EEC partners that they would honor present gas-export contracts but would not increase deliveries.[18]

Offshore exploration has proved successful in the Dutch sector of the North Sea. Most of the area is under license. Since the

[18] "Offshore Gas for Holland," *The Petroleum Economist*, February 1975, p. 45.

Groningen gas field is approaching maximum output and the demand for natural gas continues to grow, the Dutch government has now agreed to bring offshore gas discoveries into production. At the end of 1975 when the pipeline was completed, the first offshore gas flowed from the Placid fields. Half of this gas will be sold on the domestic market, through Gasunie, and half to a group of West German utility companies. According to a recent decision, the state will be entitled to a 50 percent participation in any future concession of oil and gas in the Dutch North Sea sector.[19] Total Dutch recoverable reserves of natural gas are estimated at 2,600 billion m^3.[20]

Because of the changing prospects of oil supply, the Dutch government joined the ranks of importers of natural gas. A consortium created in 1972 and including Gasunie, Ruhrgas, Distrigaz (Belgium), and Gaz de France is importing gas from the Ekofisk field in the Norwegian sector of the North Sea through the pipeline to Emden (West Germany). The share of natural gas in Dutch energy is expected to decline to 37.7 percent in 1985 and that of oil to grow to 48.3 percent. Therefore the Netherlands will become a net energy importer in the early 1980s.[21]

The present and future share of both oil and gas in the Dutch economy leaves little scope for nuclear energy. The Dutch government plans to increase installed nuclear power from the present 500 MWe to 4,000 MWe by 1985. The unit normally employed will be a light water reactor. Neratoom, a Dutch cooperative undertaking that includes the Dutch Philips Company is engaged in research with Siemens in West Germany and Belgonucléaire on breeder reactors. In addition, the Netherlands is a partner of Britain and West Germany in Urenco, the uranium enrichment venture using the centrifuge process.

A change in the Dutch oil situation occurred during the October 1973 Yom Kippur war, when the Dutch government, under the pressure of public opinion, issued a declaration stating that Egypt and Syria had unilaterally broken the truce in the Middle East. The Arab countries retaliated with an embargo on oil deliveries to the Netherlands. In November 1973, the Dutch government joined EEC members calling on Israel to end the territorial occupation it had maintained since 1967. In spite of this shift of policy, the Arab embargo

[19] "Dutch Raise State Oil and Gas Take," *The Petroleum Economist*, May 1976, p. 177.

[20] "Natural Gas in Western Europe," *The Petroleum Economist*, November 1974, p. 422.

[21] "Netherlands, Second Offshore Gas Field," *The Petroleum Economist*, April 1976, p. 145.

was pursued until May 1974. But, as a result of the manipulations of the international oil companies, the Netherlands did not suffer seriously from an oil shortage.

West Germany

The industry of West Germany, like that of Britain, was built on cheap coal, but West Germany also foresaw the future of oil. Prior to World War I the German government had achieved participation in Turkish Petroleum (later Iraq Petroleum Company—IPC), which was taken over after the war by France.

Substitution of oil for coal took place in the 1960s along very much the same pattern as in Britain. The share of coal in total primary energy consumption fell from 75 percent in 1960 to 31 percent in 1973, while the share of oil rose during the same period from 21 to 56 percent.[22] West Germany is today the largest European importer of oil. Its dependence on foreign supply of primary sources of energy amounts to 55 percent as against only 9 percent in 1960.

West Germany has modest domestic reserves of oil and natural gas, but no discoveries have been made so far in its sector of the North Sea. On the other hand, West Germany's hard and soft coal reserves are the largest in Europe, with economically recoverable reserves estimated at 40,000 megatons. There are important deposits of soft coal in the Hambacher area near the French border.[23]

For a number of years, West Germany had no other energy policy than to protect coal against oil, using a number of devices, including taxation of imported coal and a heavy discriminatory tax on oil. Only in 1968 did West Germany's coal policy enter a constructive phase with the creation of Ruhrkohle A.G., a cooperative organization producing three-fourths of the total coal output, and entitled to government subsidies. The West German government, through two partly nationalized companies, Veba-Gelsenberg (an energy conglomerate described below) and Salzgitter (a large steel concern), owns 38 percent of Ruhrkohle equity. Meanwhile, Germany remained for a long time the only large European state with no direct interest in the oil business. The U.S. and Anglo-Dutch major oil companies controlled the bulk of refineries and of the distribution network.

Finally overriding the objections of the Federal Cartel Office, the West German government sponsored the 1975 merger of two oil

[22] OECD, *Energy Balances of OECD Countries 1960-1974* (Paris), pp. 251, 264.
[23] World Energy Conference 1974, *Survey of Energy Resources* (New York: The U.S. National Committee of the W.E.C., 1974), p. 313.

companies, Veba, in which it had a 40 percent stake, and Gelsenberg. Veba now controls oil transportation and refining and petrochemicals and Gelsenberg oil and gas exploration and production.[24] The new conglomerate Veba-Gelsenberg (V.G.) controls 20 percent of West Germany's refinery capacity and owns a majority holding in Aral, an oil-distribution network comprising one-fourth of the country's gasoline stations. V.G. holds a 54 percent share in Deminex, the West German oil-exploration group, and has interests in Ruhrkohle and in Ruhrgas. West Germany has therefore shifted in recent years from a free-market economy in energy to a system in which the state exercises a major influence in the fields of coal and oil through powerful national companies.

It is against this background that one must consider the production objectives for 1985. West Germany plans to increase its total primary energy consumption from 274 million t.o.e. in 1973 to 388 million in 1985. This projection, revised downward from an estimate made in 1973, is based on an average rise in energy consumption of 3.5 percent per year corresponding to a 4 percent annual rise in the gross national product.

By the same projection, the share of hard and soft coal is to be reduced from 31 percent to 21 percent, but the quantity produced is to remain almost the same.[25] Coal is to be earmarked for steel production until after 1980 when the direct reduction process can be operated, and for conventional power stations. An important long-range research project is being launched by a subsidiary of Ruhrkohle, to provide all large cities with a district heating system by the end of the century.

The share of oil is to be reduced from 56 percent to 44 percent, while that of natural gas is to be increased from 10 percent to 18 percent.[26] West Germany, with insufficient refining capacity, is importing petroleum products from her EEC partners, especially from the Netherlands. Deminex, the government-subsidized exploration company, has bought a substantial interest in the Piper field, in the British sector of the North Sea, while West Germany has negotiated long-term contracts for natural gas with the Netherlands, with the Soviet Union, and with Norway. At the same time, an important contract for natural gas is under negotiation with Algeria. In 1975 a triangular agreement was signed with the Soviet Union and Iran to assure further deliveries of

---

[24] Veba, *Bericht über das Geschäftjahr 1975*, p. 12.

[25] Bundesministerium für Wirtschaft, *Erste Fortschreibung des Energie programms des Bundesregierung* (Bonn, 1974), p. 18.

[26] Ibid.

gas by Iran to the Soviet Union, and by the Soviet Union to West Germany. The natural gas pipeline from the Norwegian Ekofisk field to Emden has just been completed. Over the last two decades German industry has been exporting to the Middle East and to North African markets, where it has found large outlets for equipment and chemical products. It is therefore in a more favorable position than its European competitors to obtain large amounts of oil in those countries. However, the West German government has deliberately avoided concluding agreements with producer countries for the delivery of oil.

In 1973 West Germany was the second largest European producer of nuclear-generated electricity, after Britain, with a production of 4,400 MWe in ten power stations. In 1976, thirteen power stations were operating with a capacity of 6,400 MWe. Eleven more were under construction for 9,000 MWe and ten under consideration.

According to its 1974 long-range energy program, West Germany was expected to produce 50,000 MWe by 1985.[27] This goal was scaled down to 35,000 MWe by the minister of economics in a 1976 statement. A further reduction to 20,000–25,000 MWe is under consideration for ecological reasons, which have taken a political dimension. Citizens groups have demonstrated against new plants. Courts have halted work at two nuclear sites. Laender governments are reluctant to propose new sites, and the federal government has decided not to issue building permits as long as there is uncertainty about the methods and place of disposal of radioactive waste. In this situation of quasi-moratorium and with the political parties divided on the issue, it is doubtful whether the contribution of nuclear energy to total primary energy consumption will increase from 1 percent to 15 percent and account for almost some 45 percent of the total electricity generated. These figures will be reduced by more than one-half.

A major asset for the development of nucleonics was the size and high technological standards of the West German electrical and mechanical industries. Another strength was the decision to adopt first and improve later the U.S. light-water-reactor technology using enriched uranium. Thus West Germany avoided the costly experiments of Britain and France with reactors fueled with natural uranium.

West Germany has no public administration equivalent to the British Atomic Energy Authority or to the French Commissariat à l'Energie Atomique. The four-year atomic program has been jointly prepared by the Federal Ministry of Science and Technology, by the universities and other specialized research centers, and by the industry, with federal funds provided for research. In the 1960s the West

[27] Ibid., p. 42.

German nuclear industry was dominated by two companies—Siemens, which held the PWR patent from Westinghouse, and AEG, which holds the BWR patent from General Electric, with G.E. owning a 10 percent share in the capital stock of its licensee. At first Siemens and AEG were associated in Kraftwerkunion (KWU), a sales subsidiary which offers turnkey nuclear power stations to electricity producers, but AEG was forced by its financial situation to leave the subsidiary. As a result Siemens remains the major constructor of light water reactors in West Germany.[28] Siemens abandoned its Westinghouse license in 1969, and KWU is now bidding against U.S. manufacturers abroad. All orders for reactors in West Germany have been placed with West German firms, and three reactors have been sold outside the country, in Austria, in the Netherlands, and in Argentina. KWU has received an order for two reactors in Iran.

In 1975 West Germany signed an agreement with Brazil for nuclear cooperation according to which KWU was to deliver eight 1,300 MWe nuclear power stations over the next fifteen years.[29] This agreement, described in detail below, opened the world market of reactors to West German industry at the expense of its U.S. competitors, who had held a quasi-monopoly.

In the field of advanced reactors, West German industry is exploring simultaneously several avenues with various European and U.S. partners. In 1970, a high-temperature reactor using thorium (THTR) of 300 MWe was ordered for completion by 1977 in Uentrop. The companies participating in this project, Brown Boveri & Co., Hochtemperatur Reaktorbau (a joint subsidiary of General Atomics and of Brown Boveri) and Nukem, are following closely the operations of a high-temperature reactor developed by OECD in the framework of Project Dragon. The same companies are working on an improved version of the high-temperature reactor with a helium turbine.[30] West German industry is also involved in fast-breeder-reactor (FBR) research and development. In 1972 West Germany promoted a consortium with Belgium and the Netherlands to launch in Kalkar a 300 MWe sodium-cooled breeder to be completed by 1979, the West German partner being Siemens. At the same time, the major electricity producer RWE is engaged jointly with EDF in France and ENEL in

[28] "Promoting a Common Front by Energy Importers," *International Herald Tribune*, April 18, 1975, p. 18C.

[29] Dominique Verguese, "L'Allemagne fédérale construira au Brésil huit grandes centrales nucléaires en quinze ans," *Le Monde*, June 29, 1975.

[30] Bundes Ministerium für Bildung und Wissenschaft 4. Atomprogramm der Bundesrepublik Deutschland für die Jahre 1973-76 (Bonn, 1972), p. 63.

Italy in the construction of another type of fast breeder under the name of Superphénix.

West Germany has also been active in the development of the fuel cycle. Its nuclear power plants are operating with enriched uranium supplied by the United States under a 1960 long-term contract signed by the U.S. Atomic Energy Commission and by Euratom. A further agreement was signed in 1973 between West Germany and the Soviet Union for the supply of enriched uranium by 1976. But West German industry was eager to develop European-processed enriched uranium. Research conducted simultaneously in West Germany, Britain, and the Netherlands in the centrifuge method led to the constitution of a joint company, Urenco, to develop and operate centrifuge plants in Europe and to market enriched uranium. In addition, West Germany has made a breakthrough in its natural uranium supply by signing the above-mentioned agreement with Brazil. West Germany is to build a centrifuge plant in Brazil, and Brazil in return is to deliver natural uranium.

Lastly, West Germany is engaged in a cooperative venture with Britain and France for the reprocessing of nuclear fuels. A company set up to this effect under the name of United Reprocessors is operating existing plants in Britain and France and is to build a third plant in West Germany.

In spite of heavy dependence on imported oil and gas, West Germany is in a comparatively strong position in the field of energy. This strength derives from vast coal reserves and from mastery of the light-water-reactor technology in which West Germany leads Europe. West Germany is also making great strides in the technology of several types of advanced reactors, and will soon be in a position to choose the best one. The country may also become the leader in the uranium-enrichment centrifuge process.

West German progress towards energy independence could be hampered by two factors. The first is the constraint of environmental factors. In spite of the actions of the Institute on Reactor Safety, local authorities and pressure groups are opposing the construction of nuclear power plants. The second factor is the capital cost of the long-range energy program. According to a 1974 study by the Dresdener Bank, the financial needs of such a program would amount to DM 317 billion, including a 6 percent inflation rate. The Dresdener Bank's conclusion points to the existence of a DM 50 to 100 billion financing gap, and proposes the creation of an energy bank.[31] Other banking circles

31 Andrew Hargrave, "Financing Germany's Expanding Energy Program," *International Herald Tribune*, April 18, 1975, p. 24C.

recommend the development of leasing techniques for the financing of major projects.

Meanwhile West Germany maintains good relations with the oil-producing countries, where its manufactured goods and turnkey plants are in demand, because of their high technology, strict delivery dates, and efficient after-sales service. West Germany's supplies of fuel were practically unimpaired by the 1973–74 embargo on the Netherlands.

France

Between 1960 and 1973 France's energy balance sheet deteriorated sharply as a result of the explosion in oil consumption. Dependence on imports increased from 36 percent to 77 percent. The share of coal fell from 52 percent to 17 percent, that of oil rose from 30 percent to 67 percent.[32]

France has no sizable reserves of fossil fuels. Its economically recoverable reserves of coal are estimated at 458 megatons.[33] Present reserves of natural gas in the Lacq field in the southwest are on the order of 192 billion m^3.[34] Onshore oil fields are negligible.

Meanwhile, offshore exploration in the English Channel (the so-called Iroise Sea) is just starting. The uranium picture is brighter: proven reserves of U_3O_8 at a cost lower than $15.00 per pound are estimated at 37,000 tons, the largest reserves in Europe, accounting for 3.4 percent of this category of world reserves.[35] In the field of renewable energy, the best hydroelectric sites were equipped with power plants in the 1950s.

For decades France has practiced an active energy policy. Directly or indirectly, the government controls the whole energy sector. Two national oil companies, Compagnie Française des Pétroles (CFP) and Elf Erap, cover one-half of the French market. The state has a 35 percent share in the first and entirely owns the second. In 1976, at the request of the government, Elf Erap merged with its main subsidiary, Société Nationale des Pétroles d'Aquitaine, which had a large number of private shareholders. The state owns a 70 percent interest in the new company, Société Nationale Elf-Aquitaine. Under legisla-

---

32 OECD, *Energy Balances*, pp. 236, 249.

33 World Energy Conference 1974, *Survey of Energy Resources* (N.Y.: The U.S. National Committee of the W.E.C., 1974), p. 314.

34 Conseil Economique et Social, *Les Problèmes de l'énergie en France à moyen et à long terme* (Paris: Journal Officiel, 1974), p. 1309.

35 OECD, *Uranium Resources*, Table 1, p. 22.

tion of 1928, the state holds a monopoly on oil imports, the operation of which is delegated in the form of quotas to the two French and the four foreign oil companies (Esso, Mobil, Shell, and BP) under close government supervision. State-owned corporations exert a monopoly over the production of coal, with Charbonnages de France, over the production and distribution of electricity, with Electricité de France (EDF), and over natural gas, with Gaz de France (GDF). A government agency, the Commissariat à l'Energie Atomique (CEA) is in charge of military and civilian atomic research and of the development of nuclear reactors up to the prototype stage. Tariffs and investment programs of the nationalized industries are subject to government approval, and the industries are included in the overall five-year development plans.

As in other European countries, coal had to be protected in France. The growing deficits of Charbonnages de France were covered by subsidies from the national budget, in spite of which the share of coal has been declining and will probably not exceed 13 percent of primary energy resources by 1985.[36]

Of the two French oil companies, CFP was given the task of defending national interests among the major oil companies, while Elf Erap was designated to act as an agent of the French government in negotiations with the producing countries.[37] Both companies were active in exploration and production in the Algerian part of the Sahara in the 1955–65 period. However, the majority (51 percent) of their assets were nationalized by the Algerian state in 1971. Since 1965 CFP and Elf Erap have been diversifying their upstream activities in a number of "oil provinces," including the Persian Gulf, West Africa, Canada, and the North Sea, either as concessionnaires or as operators. The total crude-oil production of the two French groups in 1973 exceeds two-thirds of the domestic consumption. The government has gradually increased the share of these groups in the French market. The first exploration permits, offshore in Brittany, were awarded in 1975 south of a disputed area bordering on Britain's Celtic Sea, the dispute being subject to arbitration. Since 1973, prices of petroleum products have been adjusted upwards, but not in proportion to the higher costs of crude oil, causing the financial situation of the companies operating in France to deteriorate. In 1975 the government fixed a ceiling of F51 billion per year for oil imports and imposed restrictions on the use of fuel oil both for industry and for space

[36] Ministère de l'Industrie et de la Recherche, *Conseil Central de Planification* (Paris, February 3, 1975), p. 3.

[37] Prodi and Clô, "Europe," p. 96.

heating. Actual expenditure for 1975 was F41 billion. In 1974, domestic consumption was reduced 7.8 percent from that of 1973.[38] The French government now hopes that oil consumption will not exceed 40 percent of total energy consumption by 1985.[39]

Consumption of natural gas did not grow as fast in France as in other countries because no excise tax was charged on fuel oil. Tariffs for gas have been belatedly adjusted upwards and remain lower than costs. Meanwhile, the production of the Lacq field has reached its plateau, and a number of import contracts have been signed with Algeria, the Netherlands, the Soviet Union, and Norway, most of them since 1971. France is interested in the pipeline that is to carry gas from Ekofisk to West Germany, and is also considering a project for a pipeline for Algerian gas via Spain. Because of its active import policy, France expects the share of natural gas in its total primary energy consumption to rise from 8 percent in 1973 to 15.5 percent in 1985.[40]

Prices of electricity increased in somewhat the same pattern as petroleum product prices, though a much tougher government policy might have been expected, considering the enormous prospective investments in the field of nuclear energy. France has practically no alternative to an ambitious nuclear program, if the country is to reduce its dependence on imported oil. Nuclear energy accounted in 1973 for 1.8 percent of total primary energy consumption and for 8 percent of electricity production. Forecasts for 1985 would push nuclear energy to 23.7 percent of the total primary energy consumption and to 72 percent of electricity production. Electricity from nuclear power plants amounts at present to 3,000 MWe. An accelerated program adopted by the government in 1974 aimed at a 1985 figure in the 45,000–50,000 MWe range. However, this program was considered too ambitious in consideration of the volume of investments—in the range of F80 billion at 1974 prices—and of the capacity and structure of the French industry. Furthermore, the prospect of building some fifty nuclear power stations over the next decade has raised protests from environmentalists. The program was therefore revised downwards. Firm orders were limited to power stations of a total 12,000 MWe capacity, to be launched in 1976–77, and of 5,000 MWe capacity, to be launched in 1978.[41] These figures do not include

---

38 "Can France Win Its Oil Battle?" *The Petroleum Economist*, April 1975, p. 127.

39 *Conseil Central de Planification*, p. 3.

40 Ibid.

41 "Faits et travaux du mois," *Revue de l'énergie*, May 1976, p. 285.

capacity for exports. The government will periodicaliy review the situation.

The first French-built reactors used on a commercial basis, beginning in 1963, were of the natural uranium graphite-gas type, since France possessed uranium ore reserves. The operation of the reactors was satisfactory, but they could not compete in price with American enriched uranium PWR and BWR types. Therefore a decision to put out tenders for light water reactors was made in 1969. In entering this field, EDF already had the experience of two Franco-Belgian PWR joint operations.

The major industrial group engaged in this nuclear venture was Framatone. Its capital stock was controlled up to 51 percent by Creusot-Loire (under the control of the Belgian Empain group) and to 45 percent by Westinghouse, using the PWR licenses. Compagnie Générale d'Electricité (CGE), a licensee of General Electric for the BWR version, also was assured that it would have a share in the French reactor market on the basis of the 1974 program. Curtailment of the nuclear program in 1975 led to the conclusion that two companies could not remain profitable on a relatively small market. As a result, CGE and Compagnie Electro-Mécanique (CEM) will supply the turbo-alternators, while the engineering and the supply of the boilers and of the main components will be the responsibility of Framatone. Following negotiations in 1975 sponsored by the French government, the Commissariat à l'Energie Atomique (CEA) purchased 30 percent of the capital stock of Framatone from Westinghouse, which will be paid in natural uranium. The three concerns will join forces in research and development for the improvement of the PWR version until 1982, when Westinghouse will sell its 15 percent remaining stock to Creusot-Loire. Framatone has sold two PWR reactors to Belgium and two to South Africa, and undertaken negotiations to sell two reactors to Iran.

In the field of advanced reactors, CEA has devoted most of its research and development activities to breeders. Phénix, the first experimental breeder reactor, with a capacity of 250 MWe, became operational in 1974. In order to compete in the European market, a company, NERSA, was formed between EDF and its West German and Italian counterparts, ENEL and RWE, with the aim of building in the Rhone Valley a 1,200 MWe commercial size breeder reactor called Superphénix.[42] The decision to launch the project was made in 1976, with the works scheduled to be completed by 1979–80. At

[42] Georges Vendryes, "Aspects economiques et industriels de la politique électronucléaire de la France," *Défense Nationale*, March 1975, p. 28.

the same time, an industrial company, Novatome, was created in 1976 for the construction and sales of breeder reactors, with Creusot-Loire and CEA as shareholders, and Alsthom to join at a later stage.[43]

The CEA has invested less heavily in the high-temperature reactors, but it cooperates in research and development with General Atomics. In 1976 France and West Germany decided to pool their resources in the field of advanced reactors (see Chapter 4).

The CEA has been very active in providing resources and technology for all phases of the fuel cycle. It is in the unique position of having built a vast industrial complex covering prospecting, mining, enrichment, production of fuel elements, and reprocessing. Prospecting and production of uranium ore have been launched not only in metropolitan France but also in two French-speaking African states, Niger and Gabon. The companies under CEA's supervision control about 15 percent of the natural uranium world market.

The CEA also promoted Eurodif, a large uranium enrichment plant operating by the gaseous-diffusion process. It had gained experience in this technology in its Pierrelatte plant, which manufactured highly enriched uranium for military use. The objective was to make Western Europe independent of U.S. deliveries, soon to become in short supply because of enlarged nuclear programs in both Europe and the United States. The publicly owned Eurodif company, officially set up in 1972, had as its original partners (after Sweden opted out) France, through the CEA, with 52.8 percent, Belgium and Spain with 11 percent each, and Italy with 25.2 percent. In 1973 Iran purchased a 10 percent share through a Franco-Iranian subsidiary. Ground was broken for the plant, located at Tricastin, near Pierrelatte, in 1974, with production scheduled to begin at the end of 1978. European needs for enriched uranium in the next decade are such that the Eurodif venture should be viewed as complementary to, rather than competitive with, the future Urenco centrifuge plant operated by West Germany, Britain, and the Netherlands.[44] In 1971, the CEA, the West German industrial group KEWA, and Britain Nuclear Fuel Ltd (BNFL) joined forces in establishing United Reprocessors, to avoid duplication of investment and to market jointly their present and future capacity for the reprocessing of nuclear fuels—including plutonium produced by light water reactors—and for the disposal of radioactive waste.

The French government, in reassessing its energy policy in 1975, based its aims on a zero growth in primary energy consumption in

[43] "Faits et travaux du mois," May 1976, p. 291.

[44] Vendryes, "Politique électro-nucléaire de la France," p. 35.

1975 and a 3 percent growth per annum during the 1976–85 period. A consultative commission in charge of advising on the French government's Seventh Development Plan in the field of energy has expressed some doubts on the compatibility of moderate energy-consumption growth with the ambitious goal of a 5.5 percent growth in the gross domestic product. The commission recommended stronger energy-saving incentives, a price system reflecting the cost of energy supplies, and the launching of a public information campaign.

Meanwhile, the French manufacturing industry, with the strong backing of the government, has improved its position in the oil-producing countries. Important bilateral agreements were concluded in 1974, notably with Iraq, Iran, and Libya, and several industrial projects are under way. The Iranian government has signed a letter of intent for the purchase of two nuclear power plants with a capacity of 900 MWe each.

The successful conclusion of these agreements has been aided by France's firm stand since 1967 in favor of the Arab cause in the Middle East. As a result of a French initiative, EEC member countries on November 6, 1973, adopted an interpretation of UN Resolution No. 242 directing Israel to end its occupation of the Arab lands seized in 1967. Since then, France has endeavored to organize an EEC-Arab League dialogue, which at the request of the United States does not include oil or major political issues. No substantial achievements have as yet been made.

When at the Washington Conference of February 1974 the United States paved the way for setting up an organization of the industrial oil-consuming countries, France opposed the project on the ground that those countries should pursue a policy of cooperation and not of confrontation with OPEC. France did not join in the International Energy Agency set up in November 1974, but induced the EEC as a group and the United States to participate in a preliminary conference that convened in Paris in April 1975 to discuss oil problems between consumer countries—both industrialized and developing—and producer countries. The failure of the April meeting to reach its objectives did not deter France from pursuing its efforts to launch what is now called the Conference on International Economic Cooperation, the first meeting of which took place in Paris in December 1975.

Italy

Over the last thirty years Italy has built plants on its best hydroelectric sites in the Alps and exploited a number of medium-size natural gas fields in the Po Valley. In 1973 oil and gas were discovered in a deep-lying structure at Malossa, near Milan. There are also prospects for offshore oil in the Adriatic Sea. The country, however, has no coal resources. Its rapid industrialization since World War II led to an equally rapid increase in energy consumption, with the share of oil jumping from 41 percent in 1960 to 74 percent in 1973. Meanwhile external dependence on energy reached 80 percent in 1973 compared with 52 percent in 1960.[45]

The major policy goal of the Italian government has been to ensure from a variety of sources a supply of hydrocarbons at the lowest possible cost. The instrument in both the framing and the implementation of this policy was Ente Nazionale Idrocarburi (ENI), a powerful state trust operating under able leadership. ENI covers the whole field of oil and gas industries in Italy and abroad, and also of engineering and of offshore pipeline and platform construction. The group is at present engaged in a large-scale exploration program onshore and offshore in Italy and abroad. In a position of monopoly for the production, distribution, and import of natural gas, ENI has concluded supply contracts with Libya, the Netherlands, the Soviet Union, and Algeria. ENI companies are engaged in design and construction of a gas pipeline from Hassi R'mel in the Algerian Sahara to Cap Bon in Tunisia, and across the Mediterranean to Sicily and the Italian mainland. The pipeline is expected to be completed by 1978.

At first ENI was in a position of competition with the major oil companies on the domestic oil market. But a government price freeze in 1974 drove away several foreign oil companies, and ENI absorbed Shell Italiana, increasing its market share. Later BP was also forced out of business, while some U.S. oil companies are believed to be reconsidering their positions. Italy's proximity to the Middle East and Libya provides a comparative advantage in the price of crude-oil imports. ENI has built a powerful refining industry in the south, exporting petroleum products to the eastern Mediterranean area.

ENI's technological capacities have been successfully tested in the North Sea, where the group contracted to drill wells and lay a sea pipeline in the British sector, and also in Egypt, where the Suez-Mediterranean (Sumed) pipeline is being built by Italian enterprises. Since the rise in oil prices, ENI has entered into a number of bilateral agreements with oil-producing countries. The group's engineering affiliates are very active in Iraq and in Saudi Arabia, and have been granted new drilling permits in Libya. Thus Italy is forging new links

[45] OECD, *Energy Balances*, pp. 311, 324.

of partnership with its main suppliers, facilitated by the pro-Arab stand Italy has consistently taken in the Middle East conflict.

In contrast to these effective oil and gas policies, Italy's nuclear policy appears to be hesitant. The country's limited investments in nuclear research have excluded promotion of Italian reactors. Three nuclear power stations of the first generation with a total installed capacity of 1,400 MWe have been built to date, but they are of three different types—BWR, PWR, and graphite gas (British Magnox). This results in a dissipation of effort.

In the state industry sector, Ansaldo Meccanico Nucleare operates on license from General Electric. In the private sector, two groups operate or hope to operate: Elettronucleare Italiana, holding a Westinghouse license, with Fiat as a large shareholder; and SPIN, an association of Italian and U.S. companies, including Babcock and Wilcox. In an attempt to coordinate research, an engineering company, NIRA, was created, with Fiat, ENI, and IRI (the large state-owned industrial trust) as shareholders. NIRA is mainly involved in the study and design of fast-breeder reactors. NIRA has joined ENEL, a monopoly controlling production and distribution of electricity, in the breeder reactor project Superphénix, a joint venture among French, West German, and Italian producers of electricity and constructors of nuclear reactors.

The Italian government plans to increase nuclear-generated electric power from the present 1,400 MWe to 18,000 MWe in 1985 [46] with the aim of increasing primary electricity production to 18.5 percent of total energy consumption. Two obstacles are in the way: the difficulty of financing ENEL's investment program, because of the cheap energy policy imposed by the government, and the opposition of public opinion to the building of nuclear power plants. Political difficulties may also arise. The Communist party, a winner in the 1975 and 1976 elections, has called for a larger public—as opposed to private enterprise—responsibility in the nuclear industry, and opposes the overwhelming dependence, over the next few years, on U.S.-designed reactors (PWR and BWR). In sum, Italy's massive investment program has still not been made final as to location, financing, and type of plants.

Sweden

In 1973 almost 60 percent of Sweden's energy consumption relied on imported oil. Primary hydroelectricity provided 30 percent of energy

[46] Assembly of Western European Union, *Peaceful Uses of Nuclear Energy*, Report of Mr. Osborn, Annex II, Paris 1974, p. 30.

needs, and coal and wood the remainder. Oil consumption per capita in Sweden is the highest in the world, with the country's dependence on foreign supply amounting to 62 percent.

Sweden's imports of crude oil are divided between the Middle East, Africa, and Venezuela, those of petroleum products between Western and Eastern Europe. Imports are in the hands of the large multinational corporations and two Swedish companies which account for 25 percent of the market: OK, a unique company owned by thirty cooperative associations, and Nynäs. Approximately 40 percent of oil products are refined in Sweden. A state oil company, Svenska Petroleum AB, formed in 1975 as a subsidiary of the state mining concern LK AB, will be given the resources to develop as an integrated oil company.[47]

Up to the present, Sweden has been unsuccessful in prospecting the Baltic. The international legal situation of the continental shelf has been settled with Finland but remains unsettled with the Soviet Union and Poland. Sweden's two prospecting companies, 50 percent state-owned, have encountered difficulties, notably the failure of one of them, Petroswede, to enter the North Sea market. Recent trade agreements with Algeria, Libya, and Iran have not so far led to specific contracts for oil imports.

Sweden has built its own nuclear plants, and had five nuclear power stations with a total capacity of 2,500 MWe in operation at the end of 1974. The main constructor is Asea-Atom, owned half by the state and half by Asea, the largest private electrical group. This company's only export has been a turnkey order for a Finnish nuclear power plant. Six power plants are under construction in Sweden either by Asea-Atom or by its competitor, a U.S.-Swedish group headed by Westinghouse, and orders for two more reactors have recently been placed. With a total of thirteen reactors, Swedish nuclear-generated electricity production is expected to supply 12 percent of the primary energy demand by 1985.[48] The government also intends to install more hydropower plants.

An original feature of the Swedish approach to energy policy has been study groups organized by voluntary adult education associations involving tens of thousands of people. A definite majority of these discussion groups supported the aim, now government policy, of reducing the rate of growth in energy consumption from approxi-

---

47 "Sweden, Larger State Share Proposed," *The Petroleum Economist*, May 1976, p. 187.

48 "Sweden Opts for Low Energy Use," *The Petroleum Economist*, May 1975, p. 169.

mately 4.5 percent per annum over the last fifteen years to an average of 2 percent between 1973 and 1985. The Socialist government gave serious consideration to the possibility of leveling off energy growth by about 1990.[49] When the Socialists were defeated in the September 1976 elections, nuclear power was a major issue, but the "bourgeois" parties were divided on whether nuclear expansion should be halted.

Together with "planned resource management under popular control," international partnership is a goal of Swedish energy policy. In 1974 Sweden joined the International Energy Agency, which, in the official view, does not impinge on the country's traditional neutrality.

---

[49] Address by Prime Minister Olaf Palme before the Executive Committee of the Social Democratic Labour Party, Sundsvall, February 1, 1975.

EUROPEAN CRUDE OIL PIPELINES

— Crude oil pipelines in operation

······· Crude oil pipelines in construction

▲ Oil field

Source: Reproduced, with permission, from 1975 activities report of Union des Chambres Syndicales de L'industrie du Petrole, Paris, 1975.

4

OPTIMUM USE OF DOMESTIC RESOURCES

The national energy policies described in Chapter 3 diverge in many respects: each state has adjusted its policy to its natural resources, its technological ability, its capital availability, and its political leanings.

In order to assess the global effects of the various national efforts, the country-by-country approach must be complemented by a branch approach and an analysis of the present situation and prospects of Western Europe in oil, gas, and nuclear-generated electricity. It will then be possible to formulate a judgment on Europe's use of its domestic resources. Are European countries taking steps, especially after the oil crisis, to maximize these resources for the benefit of all, or are nationalistic attitudes prevailing?

Prior to reviewing the main sources of energy, let us look at the interaction among the major contributors to national energy policies. They comprise not only states and large companies engaged in the energy field but also a complex network of international organizations.

The Actors

All European states, even those traditionally liberal, are now engaged in active energy policies. All industry uses energy, and sudden changes in the relative prices of energy effect structural changes in industrial activity and in employment, leading inevitably to state intervention and also to a degree of cooperation between states. National companies become the instruments of state intervention, and multinational companies are forced to adjust to national policy goals. International organizations provide a forum in which the highly interdependent industrialized states debate the merits of their national policies, develop convergent courses of action, and sometimes reach collective decisions. These organizations are a feature of the European geopolitical scene and are now deeply involved in energy matters, with overlapping fields of activity.

Since 1965, the Commission of the European Communities has been responsible for the administration of three distinct organizations: the European Coal and Steel Community, the European Economic Community, and Euratom. It is empowered to submit proposals to the Council of Ministers of the EEC concerning common energy policy goals that cover all sources of energy. But the lessons of the early 1960s in the field of coal have not led to a more constructive attitude by member countries in the 1970s in the field of oil. No common European policy has been adopted—only guidelines that are not binding for governments.

The oil committee of the Organization for Economic Cooperation and Development (OECD) has been less a policy-oriented body, than a forum for gathering information among the representatives of the United States, Japan, and the Western European countries. At the time of the oil crisis, the OECD became the main center for energy policy discussions because of the weight of the United States as a superpower, of its political influence and economic interests in the Middle East, and of its relative self-sufficiency in primary energy compared with that of its partners. In 1974 the United States quite naturally chose the OECD as the parent organization for the International Energy Agency (IEA), which was to exert operational functions. (IEA membership is as follows: the United States, Canada, Japan, Austria, Belgium, Denmark, Greece, Ireland, Italy, Luxembourg, the Netherlands, Spain, Sweden, Switzerland, Turkey, the United Kingdom, and West Germany.)

Some duplication between the EEC and the OECD-IEA complex became unavoidable. Europeans—specifically the EEC Nine—needed to debate among themselves and at the same time to pursue a permanent dialogue with the United States. Since the European Free Trade Area (EFTA) has no responsibility in industry or energy, OECD and IEA are appropriate forums for exchanges of views between EEC and EFTA member countries. France has not joined the IEA, however, and Norway is only an associate member. Lastly, financial and monetary problems pertaining to energy are being discussed almost simultaneously in the EEC, in the OECD, and in the IMF.

It becomes apparent in the field of energy, as in many other economic fields, that the European picture is the sum of several purely national policies, rather than an effective integrated common policy.

The international organizations operate in a gray area where individual states attempt to reconcile major policy differences and act in common only on issues after a consensus has been reached.

Against this background one must review the ways in which Europeans have managed their domestic resources in the main branches of energy, and how they are reacting to the present challenge. Occasional reference will be made to external resources, to be analyzed in Chapter 5 where the same actors will be considered in a wider context.

Coal

Coal is basically a domestic source of energy. Its decline in Europe is due to a number of factors. One of them, which has been mentioned, is price competition with oil, at a time when higher real wages were handicapping a labor-intensive industry. A second factor is a rigidly controlled market. On the Continent ECSC enforced the principles of market disclosure and nondiscrimination in sales to consumers, while coal prices in various countries were held down without regard to production costs.[1] In contrast, oil companies enjoyed a much greater flexibility in their operations. A third factor is depletion of reserves. National plans for an orderly withdrawal of coal-mining activities while recycling miners to other jobs were implemented throughout Europe, the pace being adjusted to the anticipated depletion and decreasing productivity of the mines.

However, the 1973 crisis changed government attitudes in regard to coal. Its contribution to energy needs could no longer be neglected. The share of coal in the total primary energy supply of Europe is expected to be slightly higher in 1985 than that of nuclear energy (14.8 percent as against 13.5 percent, see Table 2). To achieve that goal, coal production must be maintained slightly above the present level, and coal imports must be increased. If coal production is to be maintained, it must be assured of regular outlets at competitive prices.

The natural outlets for coal are coke production and coal-fired power stations, which now account for 80 percent of EEC consumption. The demand for coking coal will rise moderately in the next decade, in conjunction with the needs of the iron and steel industry. Deliveries to power stations will increase much more rapidly—by 25 percent in 1985—if less use is made of petroleum products as boiler fuels, and if construction of new coal-fired power stations and improvement of the existing ones are undertaken. Coal could then satisfy part

[1] Prodi and Clô, "Europe," p. 93.

of the projected growth rate of conventional electricity generation with an acceptable environmental impact.

Coking coal sold to the steel industry draws Community financial aids based on long-term contracts. The Brussels Commission recommends that a similar system of support be established for sales of coal to the power stations.[2] Any support system, however, must take into consideration the relative prices of coal and oil. Increase in productivity in the labor-intensive EEC coal industry does not exceed 4 percent per annum. If the increase in real wages of the miners is higher than that figure, the competitive position of European coal in relation to other sources of energy will deteriorate.

On the supply side, according to Community experts, coal production could be increased by 40 million tons per year in Britain, mainly in the Midlands and in Yorkshire. At the same time, an increase of 10–15 million tons could be achieved in some of the West German coal fields.[3] Germany also plans to raise the production of lignite in the Rhineland by the opening of new basins. In Belgium the Campine basin, producing coking coal, is to be maintained, but the Borinage basin will be closed in the early 1980s. In France, decline in production has been slowed down, but the North basin is scheduled to be closed in the late 1980s. Outside the Community there are good prospects for the development of coal production in Turkey and of coal and lignite in Spain.

Increasing coal imports will not be easy if Europe relies on its traditional suppliers. The United States is unwilling to engage in any long-term export commitment with EEC because of domestic uncertainties over the levels of production and the future availability of transportation facilities. Poland will not increase its deliveries to EEC because its surplus production is earmarked for domestic consumption and for sale to Eastern Europe. However, supply contracts have been signed with South Africa, whose coal is easy to mine and transport. Australia, a potential supplier for Europe, is rapidly increasing production and loading capacity.

In the field of new technology, open-air coal gasification has proved inappropriate for Europe because of its high rate of energy consumption. New processes for underground gasification are in the research and development stage. A research unit on coal has been set up in the International Energy Agency. Its purpose is to expand

[2] Commission of the European Communities, *Medium Term Guidelines for Coal 1975-1985*, Com (74) 1960 final, Brussels 1974, pp. 5, 9, 23.

[3] OECD, Groupe de Travail Mixte sur la Réappréciation du Rôle du Charbon, report of M. François Long (CEE), 1975, p. 9.

the data base on world-coal reserves, to compare the economics of coal recovery, and to accelerate the use of more advanced technology both in mining and in utilization. Germany, meanwhile, is engaged in a long-term project of producing high-class gas from coal with the aid of nuclear-process heat deriving from a high temperature reactor.[4]

Oil and Gas

In Europe as elsewhere, oil is the biggest industry but until the discoveries in the North Sea, Europe did not have an integrated capability from the well to the pump in its own territory. The main activities were downstream; they are now moving upstream. This development, together with the wide fluctuation in energy prices, has altered the respective roles of the industry's participants, leading in turn to a consideration of the relations between the oil companies and the states, the impact and prospects of North Sea oil and gas, and interstate relations.

The Oil Companies and the States. In the period between the wars, the European oil industry was for all practical purposes in the hands of multinational companies. This oligopoly was backed by the political influence of the United States and Britain, the home countries of the companies whose business was the production and transportation of oil to Europe. The only other company with a token presence in the Middle East was the Compagnie Française des Pétroles.

Table 8, compiled from *Fortune* magazine, gives the ranking by sales in 1975 of oil companies operating in Western Europe. There is a very large gap between the sales of the two European majors, Royal Dutch Shell and BP, and those of the other European-based companies. European subsidiaries of the U.S. majors are in strong positions in several European countries, and it is interesting to observe that the four Esso (Exxon) companies in West Germany, France, Britain, and the Netherlands together have a turnover of $9.3 billion, just above that of CFP. The dominant position of British and U.S. companies is unquestionable in upstream operations throughout Europe. It derives to a large extent from the hold these companies have had, through their concessions, on the largest and cheapest oil reserves in the world.

Most of the national oil companies, with the notable exceptions of CFP and Petrofina, were created after World War II. Several of

[4] "Synthetic Natural Gas Technology on Brink of Success," *The Petroleum Economist*, August 1976, p. 311.

Table 8

LARGEST OIL COMPANIES IN WESTERN EUROPE
(ranked by sales in 1975)

World Rank [a]	Rank Outside U.S. [b]	Companies	Country	Sales ($ million)
		European-Based Companies		
3	1	Royal Dutch Shell	Netherlands, Britain	32,105
8	3	British Petroleum (BP)	Britain	17,285
18	6	Cie Française des Pétroles (CFP)	France	9,145
22	10	ENI [c]	Italy	8,334
35	18	Elf-Aquitaine [c]	France	7,165
	47	Petrofina	Belgium	3,972
	100	Veba Chemie [e]	West Germany	2,050
	101	Burmah Oil	Britain	2,032
	123	Enpetrol	Spain	1,724
	133	Gelsenberg [d] [e]	West Germany	1,668
	151	Ca. Esp. de Petroleos (CEPSA)	Spain	1,393
	167	Soc. Ind. Belge des Pétroles (SIPB)	Belgium	1,286
	187	OMV [c]	Austria	1,139
	200	Neste [c]	Finland	1,060
	204	Türkiye Petrolieri [c]	Turkey	1,038
		European Subsidiaries of Major American Companies		
	58	Esso	West Germany	3,326
	77	Esso	France	2,644
	81	Esso Petroleum	Britain	2,578
	144	Deutsche Texaco	West Germany	1,486
	160	Mobil Oil	West Germany	1,328
	186	Texaco	Britain	1,141
	268	Mobil Oil Française	France	785
	287	Esso Nederland	Netherlands	741
	297	Mobil Oil	Britain	717

Table 8 (continued)

	Major American Companies	
1	Exxon	44,864
4	Texaco	24,507
6	Mobil Oil	20,620
9	Standard Oil of California	16,822
12	Gulf Oil	14,268

[a] Rank among the fifty largest industrial companies in the world.

[b] Rank among the 500 largest industrial corporations outside the U.S.

[c] Government-owned.

[d] Parent company only.

[e] Veba Chemie and Gelsenberg merged in 1975.

Source: *Fortune*, August 1976, The *Fortune* Directory of the 500 largest industrial corporations outside the U.S., p. 231; the fifty largest industrial companies in the world, p. 243.

them are state-owned, such as ENI, Elf-Aquitaine, OMV, Neste, and Türkiye Petrolieri.

Table 8 does not reflect the real weight of Gelsenberg, since the sales figures in the table concern the parent company only. Furthermore Veba and Gelsenberg merged in 1975, giving the German consortium a rank between Elf-Aquitaine and Petrofina. Two potentially important state-owned companies are not in this table because they became operational quite recently: the British National Oil Company (BNOC) and Statoil in Norway (see Chapter 3).

While contributing to the expansion of the market through an aggressive price strategy, the major companies had to build adequate transportation and refining capacities. With the exception of a few joint ventures, they generally operated by themselves in refining, but always associated among themselves and with national companies in ground transportation.

The massive growth of consumption in the 1960s was accompanied by a parallel growth in refinery capacity, much greater in Europe than in the United States (though much slower than in Japan).

Rotterdam is the largest refining center on the North Sea, supplying north Germany. Inland refineries were built to supplement coastal refineries, as industrial production expanded in northern Italy, eastern France, and southern and central Germany. An oil-pipeline network was laid in Central Europe to supply these refineries with crude oil from the ports of Marseilles and Genoa. Thus tankers saved transportation costs around the coasts of the Iberian

Table 9

REFINERY CAPACITY

(million tons per year)

	1960	1972	1974
Western Europe	231	853	973
United States	474	670	745
Japan	33	220	254

Sources: OECD, *Oil, the Present Situation and Future Trends*, Paris, 1973, p. 105; *BP Statistical Review of the Oil Industry*, 1972, 1974.

peninsula and France for oil shipped from the eastern Mediterranean and Algeria (see map on page 66).

Except for the two major European oil companies, European national companies were short of oil reserves and consequently of crude supplies. They operated in narrow consumer markets and could not benefit fully from economies of scale. Their financial situation, compared with that of the major companies, was much less favorable in terms of volume of owned assets, profitability, cash flow, and leverage. Observing that the U.S. companies benefited at home from significant tax advantages which helped them expand abroad, specifically into Europe, European companies asked EEC authorities in 1967 to help establish fair competition in the Common Market, but no action was taken.

The same companies felt the need to concert their policies as they suffered from the reduction in crude oil imports and petroleum product sales in 1974 and 1975. In the circumstance, their recent investments in refining and in tanker fleet proved excessive. The companies were overequipped at a time when costs were increasing and when governments, with the exception of that of West Germany, held down the prices of products. Thus the European-based companies were operating at a loss and turned again to the European Commission for help in 1976.

Excise taxes on petroleum products varied widely from country to country. So did the regulation of the national oil markets, ranging from almost complete freedom in northern Europe—national and foreign oil companies being treated on the same basis—to strict regulation of market shares and containment of the major oil companies in France. Mediterranean countries also followed interventionist policies, with relationships between government and companies varying accordingly. France and Italy used state-controlled companies to foster cooperation with the oil-producing countries, and encouraged them in geographical diversification. In West Germany, the major companies controlled approximately 75 percent of the market in the late 1960s. Only then did the government endeavor to set up a consortium of West German companies, some with state participation, culminating in 1975 with the Veba Gelsenberg merger.

In each European country, oil companies producing natural gas sold their product to monopoly enterprises under state or regional control in each country. The gas was distributed and the markets regulated on a national basis. Intra-European trade began with exports from the Netherlands to West Germany, Belgium, and France, and has expanded with exports from the Norwegian sector of the North Sea to the same three countries. Meanwhile, the Soviet Union exports gas to Austria, West Germany, Italy, and France, joined by Iran through a deal with the Soviet Union. Libya and Algeria send supplies to Italy, France, and Spain.

Ground transportation for natural gas is provided by pipelines. Major lines distribute national gas and deliver imported gas to the large urban and industrial areas. Transportation by sea takes place in the form of liquid natural gas (LNG). But the cost of liquefaction plants and specialized tankers has stimulated the construction of sea pipelines in the North Sea, a measure also under consideration in the Mediterranean. A project for supplying gas by pipeline from Algeria to Sicily and Italy is well under way, with the laying of submarine lines due to start in 1977. A projected pipeline under the Straits of Gibraltar for the delivery of Algerian gas to Spain and France is under feasibility study.[5]

In a climate of growing oil nationalism, European governments have never seriously considered the creation of a large multinational "European" company through the pooling of assets of national companies.

North Sea Developments. Exploration in the North Sea started after the discovery in 1959 of the major Groningen gas field in the Netherlands. The first offshore finds of natural gas were made in the southern basin of the British sector in the 1965–69 period. The major discoveries of oil, often with associated gas, have taken place since 1970, most of them along the border line between the British and the Norwegian sectors, on a north-south axis from Statfjord and Brent to Ekofisk. The map on page 66 shows the oil pipelines laid so far in the North Sea.

[5] "Gas Pipelines to Europe," *The Petroleum Economist*, June 1976, p. 226.

Production opportunities and constraints derive from the volume and location of reserves, which in turn bear on technical costs and financing. But the pace of production in the future will depend very much on government decisions.

Figures for oil and gas reserves in the North Sea vary from year to year in relation to the progress of exploration and production. In a medium-term perspective, not only the proven but also the recoverable reserves must be taken into consideration. A 1975 survey covering the whole offshore area from the Celtic Sea, south of the British Isles, to the Barents Sea, north of Norway, quotes a figure of 4 billion t.o.e. for proven reserves, plus 1.7 billion t.o.e. for possible recoverable reserves at an underwater depth of less than 200 meters. Deeper reserves are not likely to be tapped for production before 1985, for technological reasons.[6] These estimates for recoverable reserves are less optimistic than those of a BP expert who expects a recovery of some 5 billion t.o.e. mostly in the northern basin of the North Sea.[7] In the controversial but even more optimistic opinion of European oil expert Peter Odell, only one-tenth of the North Sea area has been explored, and oil reserves will amount to some 15 billion tons.[8]

The North Sea is undoubtedly a high cost area, with difficult physical conditions—tide and wind. Metal platforms can cope with fields at a maximum water depth of 120 meters. Concrete platforms can cope with fields at a depth of 120-200 meters but require twice as much capital outlay. A recent estimate of the technical cost of a medium-sized field (200 million t.o.e. of recoverable reserves) with a water depth of 100 meters is $4.50 per barrel at 1974 prices. The cost soars to $6.26 for a medium-sized field (80 million tons), and to $10.48 for a small one (30 million tons).[9] A typical field of less than 500 million barrels of reserves (68 million tons) is not now a viable commercial enterprise unless special tax provisions offset technical handicaps.[10]

Oil companies have traditionally financed exploration and production costs from oil-generated funds. But OPEC-set prices for crude oil, control of the price of petroleum products in consumer countries, and cost increases in the inhospitable North Sea have led to a

[6] Elf Erap, *Monthly Information Bulletin* (Paris, May 25, 1975), pp. 5, 8.

[7] F. Howitt, "North Sea in a World Context," *Nature*, vol. 249 (June 21, 1974), p. 700.

[8] Peter R. Odell, interview in *Newsweek*, July 28, 1975. See also Peter R. Odell and Kenneth E. Rosing, *The North Sea Oil Province* (London: Kogan Page, 1975).

[9] Elf Erap, *Monthly Information Bulletin* (Paris, May 25, 1975), pp. 5, 8.

[10] Judith Gurney, "Energie aus der Nordsee, die bedeutung des Nordsee Öls und Erdgasse für West Europa," *Europa Archiv*, September 25, 1975, p. 562.

decline in financing from internal cash flow. Companies meet exploration costs from their cash flow, but must borrow from banks for the development of commercial operations, the loans being repaid out of the proceeds of production. The U.S. mortgage system allows banks to take oil in the ground as security. The Norwegian government has introduced this system, but so far the British government has not. Banks hesitated to invest as long as doubts remained about government tax and participation policies and about the adoption of a floor price.[11] These uncertainties have now been dispelled.

Governments have used their discretionary powers in license allocation, giving preferential treatment to national oil companies, nationalized industries, and foreign oil companies operating on the national market. Governments have also insisted that oil and gas be landed on and substantially processed in their own territories. Thus the British Gas Corporation is the sole importer of natural gas produced in the British sector. Similarly, pipelines from Ekofisk and Frigg go to Britain and Scotland, but tankers bring the oil back to Norwegian refineries. Governments will continue to control the issue of refinery licenses, weighing the advantage of exporting refined products rather than crude oil to Continental Europe.

Both the British and the Norwegian governments have been tightening their tax terms. In the British sector, the total government take, at the present rate of 45 percent of oil revenue for the petroleum revenue tax, will amount to about 70 percent over a period of years. Furthermore, because of the so-called ring fence device, oil companies cannot subtract the costs of operations outside the North Sea from profits of their operations in these waters, a restriction that runs counter to the integrated strategy of the companies.[12] Tax reliefs may be granted for marginally economic fields. In Norway, the government take, excluding taxes on dividends, will amount to an average of between 57 and 68 percent of earnings during the life of the field. Here the ring fence device is also enforced.[13]

Partial or total state ownership has made great strides in conjunction with the exploitation of North Sea resources. Both Norway and Britain have adopted the French pattern of placing a fully state-owned oil company over a private company in which the state holds a large participation. Statoil and the British National Oil Company (BNOC) supplement respectively Norsk Hydro and BP, just as Elf

[11] Louis Turner, "State and Commercial Interests in North Sea Oil and Gas," p. 106, and Ewan Brown, "Finance for the North Sea," in *Political Implications*, p. 119.

[12] Turner, "State and Commercial Interests," p. 102.

[13] "Norway, New Taxation Rules," *The Petroleum Economist*, March 1975, p. 97.

Erap supplements Compagnie Française des Pétroles (CFP). Both Statoil and BNOC are entitled to purchase majority participation in any new commercial oil field, while BNOC can also negotiate the purchase of shares in existing ventures, and is exempt from the Petroleum Revenue Tax. The British and Norwegian governments, under the principle of no concession without participation, are moving towards the French concept of "delegated monopoly," and are able to exert a strict control over their national oil markets.[14]

Both governments have published their production goals for 1985: 100–150 million tons of oil and 44–52 million t.o.e. of gas on the British side, 75 million t.o.e. on the Norwegian side (see Chapter 3). Taking a median figure for Britain and adding the two governments' estimates, a total of 250 million t.o.e. will be brought on the market in 1985. This figure falls short of the OECD projections for offshore indigenous production of oil and gas in Europe by 1985, a total of 444 million t.o.e., assuming a price of oil of $9.00 per barrel at 1972 prices.[15] The French company Elf Erap foresees a total production of 350 million t.o.e. in 1985,[16] while Peter Odell projects a figure of 300 million tons of oil in 1983, not including natural gas, for a possible total of 500 million t.o.e.[17]

Interstate Relations. Interstate relations in oil and gas take place on the one hand in connection with the North Sea operations and on the other hand in connection with policy deliberations in the EEC and in the IEA, where the oil supply of Europe is considered in a broader context.

The legal framework for exploitation of the North Sea derives mainly from the 1958 UN Convention on the Continental Shelf. Delimitation of the border states has been effected by a number of bilateral agreements, leaving each state with exclusive rights in its sector. Regional conferences have filled some of the gaps of international law on questions of jurisdiction, pollution prevention, and compensation, and safety standards of installation and operation.[18] Some matters remain unsettled, such as the final definition of the border between Britain and France in the English Channel and on the western approaches. The division of the continental shelf has not

[14] Turner, "State and Commercial Interests," p. 105.

[15] OECD, *Energy Prospects*, vol. 1, Table 2-6, p. 51.

[16] Elf Erap, *Monthly Information Bulletin* (Paris, May 25, 1975), p. 9.

[17] Odell, *Newsweek*, July 28, 1975.

[18] Patricia Birnie, "The Legal Background," in *Political Implications*, pp. 20, 27-32.

eliminated the risk of disputes over such matters as the distribution of reserves within fields that cut across border lines, and constitutes a step backward from what might have been a collective exploitation.[19]

Except for its Norwegian sector, the North Sea falls within European Community boundaries. According to a 1970 EEC memorandum, the continental shelf is considered the territory of a member country. Oil and gas belong to the country that has extracted them. But the free movement of goods must be guaranteed between member states. Export restrictions are forbidden as discriminatory trade practices.[20] Within this context, special agreements have been made between producer states (Britain and Norway) and between producer and consumer states for the construction of oil and gas pipelines in the North Sea and on the Continent.

In 1972 North Sea development prospects persuaded the Norwegian electorate not to join the European Community so that the country would have the new resource entirely at its disposal. The present institutional setup is not conducive to common European decisions because Norway remains outside EEC, and the IEA is set up for international energy cooperation on a much wider basis. Both the EEC and the IEA have been concerned with stockpiling and emergency sharing of resources, with varying results. A stockpile for ninety days of oil consumption was established in the EEC countries as of January 1, 1975. The non-EEC members of the IEA have not yet implemented a similar measure. The sharing of resources offers better prospects of being realized in the wider setting of IEA, with contributions from North American sources. It is in the IEA that a sharing agreement has been reached (with the participation of Norway) on the basis of the "vital needs" of member states. France has so far objected to an emergency-sharing plan within EEC which would become a part of the larger IEA scheme.[21]

Energy conservation is all the more necessary because the 10 percent price increase decided on by the OPEC countries in September 1975 imposes a supplementary burden of $5 billion on the EEC oil-import bill. A savings in consumption realized in 1974 and 1975 resulted from mild winters and depressed economic conditions. The EEC Commission proposed voluntary energy-saving targets of 20 and 30 million tons of oil in 1976 and in 1977, but no agreement had been reached in the Council of Ministers by the end of 1975.

---

19 Ian Smart, "The Impact of North Sea Oil and Gas Development on International Politics," in *Political Implications*, pp. 146-47.

20 Birnie, "The Legal Background," p. 33.

21 M. Henri Simonet, Parlement Européen, Session 1975-1976, July 8, 1975.

The optimal level of production of domestic oil and gas in Europe involves a basic policy decision. A balance must be struck between self-sufficiency and the interdependence of Europe and OPEC. Maximum relative autarky at the European level might somewhat weaken the OPEC oligopoly but would be costly in terms of investments. The EEC has little influence even on the production policy of its member states, and it has no means of inducing a nonmember state such as Norway to step up its production target, which is well below capacity. In a supply shortage, the EEC might have some influence on the condition that member countries made a financial contribution.

The matter of production level is linked to that of a minimum safeguard price, the goal of which is to protect domestic investments in the case of a decline in world oil prices. Such a minimum price was proposed to the IEA by Secretary of State Henry Kissinger. Countries without oil resources, such as Japan, Italy, Sweden, and Denmark, oppose such a move, which of course is strongly supported by Britain and Norway. The counterpart of a price safeguard in favor of European oil producers would be some sort of commitment to supply European oil importers on a nondiscriminatory basis. Whatever the final agreement, a rent will have to be paid to Britain and Norway in the form of a transfer price.

In February 1976 the IEA adopted a recommendation in favor of a floor price of $7.00 per barrel. In January of the same year the EEC Commission proposed the same floor price, to be implemented in the form of an import levy, stressing that a minimum safeguard price at this level would insure the profitability of approximately 70 to 75 percent of energy production scheduled for 1985.[22] But the Council of Ministers, meeting in March 1976, failed to agree on the floor price. France and Italy objected to the insufficiency of investment guarantees in sectors other than oil.[23]

Production targets and price guarantees must be flexible and adjusted to the situation of the market. It is therefore essential that policy makers possess adequate information on imports and exports of oil and gas, on the planning and execution of investment projects, and on the various price and cost elements for crude oil and for refining and distribution operations. In September 1974 the EEC Council of Ministers adopted a resolution to this effect.[24] Implemen-

[22] EEC, *Implementation of Energy Policy Orientations Formulated by the European Council* (Brussels, 1976), Com (76) 20, p. 5.

[23] "European Community, Disagreement on 'Floor' Price," *The Petroleum Economist*, May 1976, p. 190.

[24] EEC, *Towards a New Energy Policy Strategy* (Brussels, 1974) Com (71) 550 final.

tation measures in the form of a directive were adopted in March 1976.[25] Meanwhile, IEA authorities state that their general information system is well established, and that it provides for greater transparency of the market.

Obviously energy is an overlapping area between IEA, as the operational branch of OECD, and EEC. The two organizations have the same broad functions for emergencies and for developing long-term programs. A duality of approach is unavoidable because the Community market is a component of the world oil market. The same is true in the field of nuclear energy. The scope of the two organizations and relations between them are further discussed in Chapter 5.

Nuclear Energy

The role of nuclear energy is to provide an alternative form of electricity, which can substitute in part for electricity generated by oil- or gas-fueled power stations and decrease Europe's dependency on hydrocarbons. Furthermore, nuclear power stations can combine the production of electricity with that of heat for industrial and domestic uses, thus reducing energy waste.

Although the nuclear manufacturing industry, which is in charge of supplying reactors and components to power plants, is not nationalized, production and distribution of electricity are generally entrusted to monopolistic state-owned utility companies. In most European countries the state operates and/or finances the nuclear research and development stage, makes the basic technological choices, determines the production schedule, helps provide outlets abroad, and earmarks the external sources of investment. Political prestige and security preoccupations are further reasons for strict and continuous government supervision.

European prospects for nuclear-generated electricity must be considered from the aspects of proven reactors, advanced reactors, the fuel cycle, and the industrial structure. The national approach has prevailed in the field of proven reactors. A measure of interstate cooperation is taking place in the fields of advanced reactors and of the fuel cycle.

Proven Reactors. As mentioned earlier, Britain and France built gas cooled natural uranium reactors but abandoned this technology in the

[25] "Conseil de l'Energie de la Communauté Européenne du 25 mars 1976," *Revue de l'énergie*, May 1976, p. 276.

1960s, after failing to market their production outside their own borders. The only exceptions to this pattern were the sale of a French reactor to Spain and of one British reactor to Italy and another to Japan. Both countries were unwilling to pool their forces and were unable to resist the pressure of the U.S. multinationals, Westinghouse and General Electric, who prevailed because of their technology and sales organization. Westinghouse promoted the pressurized-water and GE the boiling-water version (PWR and BWR) of the light water reactor (LWR).

Britain was the only European country to opt out of LWR technology and engage in a limited program, with a derivative form of the heavy water reactor, fueled with enriched uranium—the "steamer." Sweden has evolved its own version of the BWR and has sold a plant to Finland.

As matters now stand, the European market is open to the U.S. light water reactor for at least the next decade. Orders are being given mostly to licensees of Westinghouse and General Electric in Europe, with one major exception. The German company Siemens, which is licensed to Westinghouse, has registered a number of orders through its sales organization, Kraftwerksunion (KWU). The French licensee of Westinghouse, Framatome, on the other hand, is still very much dependent upon its U.S. licensor and will benefit from the 1975 government decision to concentrate on a single type of reactor and to give up building BWRs. Italy is trailing in the market with three different types of reactor and no clearcut production policy.

It is difficult to assess the size of the European market for proven reactors. Previous production targets were never met because of technical obstacles and limitations in the capacity of the industry. The sharp increase in the price of oil has led several countries to adopt accelerated programs, but implementation may be hampered by the difficulty of financing the high volume of investments required. In some countries, particularly West Germany, Sweden, and Switzerland, fears about reactor safety have brought opposition from environmentalists and delays in the commissioning of nuclear plants.

Table 10 compares reactor capacity with forecasts for 1980, which are probably accurate since they combine the capacities of plants in operation and under construction. The forecasts for 1985 are much more volatile for reasons mentioned above. This third group of figures represents a hope rather than a realistic expectation.

Because of industrial, financial, and/or public opinion uncertainties, some countries have decided to revise their programs every other year. Such has been the case for France, whose industry may

Table 10

NUCLEAR POWER GROWTH ESTIMATE

(in GWe)

	1975 Actual	1980 Forecast	1985 Forecast
OECD Nations in EEC			
Belgium	1.7	3.5	9.5
Denmark	—	—	1.8
France [a]	2.3	20.4	56.0
Ireland	—	—	0.7
Italy [a]	0.6	1.4	26.4
Luxembourg	—	—	1.2
Netherlands	0.5	0.5	3.5
United Kingdom	4.8	11.1	14.7
West Germany [a]	3.2	19.1	44.6
Subtotal	13.1	56.0	158.4
Other OECD Nations in Europe			
Austria	—	0.7	3.0
Finland	—	1.5	3.9
Greece	—	—	0.6
Norway	—	—	—
Portugal	—	—	1.4
Spain	1.1	8.7	23.7
Sweden	3.2	7.4	11.3
Switzerland	1.0	3.8	8.0
Turkey	—	—	0.6
Cumulative Subtotal	18.4	78.1	210.9
Other OECD Nations			
United States	40.1	82.2	205.0
Canada	2.5	7.2	18.4
Japan	6.6	17.0	49.0
Australia, New Zealand	—	—	—
OECD Total			
High estimate	68.0	185.0	484.0
Low estimate	68.0	171.0	437.0

[a] French, Italian, and German figures for 1985 should be revised downward to 45, 18, and 20 GWe respectively (see Chapter 3).

Source: OECD, *Uranium, Resources, Production and Demand,* Paris 1975, Table 8, p. 28.

not meet the accelerated program, and for Sweden, where public opinion has resisted nuclear developments. In Germany, opposition of environmentalists resulted in the freezing of the nuclear program.

Advanced Reactors. In the medium term, the European industry is committed to two major technical options: the fast-breeder reactor (FBR) and the high-temperature reactor (HTR).

HTRs offer the advantage of harnessing waste heat at a temperature sufficient to provide cheap process heat, and it reduces dependence on cooling water.[26] This type of reactor was first used in 1964 when a 20 MWe Dragon reactor located in Britain and sponsored by the European Nuclear Energy Agency (an OECD agency) became critical, proving the technical feasibility of the HTR. Research was subsequently conducted in West Germany and in the United States through General Atomics, a joint subsidiary of Gulf Oil and Shell, which had concluded an agreement with Dragon. However, when the U.S. prototype was behind schedule, confidence in the second generation of HTRs waned, and in 1975 options taken by several utility companies were withdrawn. Gulf and Shell may well close down the line. On the other hand, Britain planned to phase out the Dragon project in 1976 because of its concentration on the Steamer reactor only. This move would be another setback for HTR development. In West Germany an AVR experimental reactor of 15 MWe became operational in 1966, and a 300 MWe plant fueled with thorium is under construction. Two large 1,160 MWe plants, scheduled for operation in the mid-1980s, require the creation of a fuel cycle built on thorium and highly enriched uranium. Meanwhile, France has entered into an agreement with General Atomics, which is still in the exploratory stage. All things considered, HTRs may have a future in the world and in Europe, but that future is still uncertain.

Fast-breeder reactors are fueled with a mixture of plutonium and uranium oxides. They are highly efficient, using at least 50 percent of the total energy from the fuel (compared with 1–2 percent for LWRs), and creating more fissionable material than they use. In spite of plutonium's high toxicity and aggressive potential, involving serious health and security problems at the commercialization stage, research is being actively pursued because the breeder offers the best prospects for the generation of electricity in the 1990s.

Three projects are under way in each of the three larger European countries. The French reactor Phénix, with a 250 MWe capacity,

---

[26] A. J. Surrey, "The Future Growth of Nuclear Power," *Energy Policy*, September-December 1973, p. 210.

became operational in 1973, a couple of years ahead of the British PFR, with the same capacity. West Germany, in association with the Benelux countries, is producing the SNR reactor, of a 300 MWe capacity, to be operational in 1980.

France is launching, jointly with West Germany and Italy, an industrial version of Phénix, called Superphénix, scheduled for a 1,200 MWe capacity and expected to be operational in 1980. Three major public utilities, EDF in France, ENEL in Italy, and RWE in West Germany are associated in this venture.

In 1976 France and West Germany agreed to cooperate closely in the field of advanced reactors. They will adopt a common research program for breeders and transfer the technology to a joint company, to be created by CEA and Novatome on the French side, and by Interatom on the West German side. This company will be the sole licensor. A similar type of cooperation is under consideration for high temperature reactors.[27]

Present U.S. policy is to hold up on a fast reactor program while the technology is being developed in Europe. There is little doubt that U.S. industry will catch up without contracting for European licenses. Breeders will become commercial when Westinghouse and General Electric are ready to sell them.

Table 11 shows the present stage of the European advanced-reactor projects.

Fuel Cycle. The fuel cycle involves the extraction of uranium ore, its conversion into uranium hexafluoride (UF^6), the enrichment of uranium in isotopic separation plants, the production of fuel elements, and the reprocessing of radiated fuels.

To meet targets of nuclear generating capacity of 160,000 MWe for EEC and of 210,000 MWe for OECD Europe, European countries must secure adequate supplies of natural and of enriched uranium. The latter has become a major prerequisite, since no more reactors fueled with natural uranium are to be built.

The production of natural uranium is likely to expand rapidly during the next two decades, until breeder reactors come to play a major role in the production of electricity and the generation of plutonium fuel. The cumulative world needs of natural uranium by 1985 are estimated at 800,000 tons, of which one-fourth will be required by Western Europe, whose reserves are insignificant except for those of France and Sweden.[28]

---

27 "Faits et travaux du mois," *Revue de l'énergie,* June 1976, p. 347.

28 EEC, *Toward a Community Nuclear Fuel Supply Policy* (Brussels, 1974), Com (74), 1963.

Table 11

ADVANCED REACTORS

	Name	Country	Fuel	Electric Output	Year of Operation
I	*HTR*				
	Dragon	U.K.	Ue	8 MWe	1964
	AVR	West Germany	Ue	15 MWe	1966
	THTR	West Germany	U/Thorium	300 MWe	1978
	VEW	West Germany	U/Thorium	1,160 MWe	1983
	GKM	West Germany	U/Thorium	1,160 MWe	1985
II	*Breeder*				
	Phénix	France	Pu 02/U02	250 MWe	1973
	PFR	U.K.	Pu 02/U02	250 MWe	1975
	SNR 300	West Germany/ Benelux	Pu 02/U02	300 MWe	1980
	Super-phénix	France/Italy/ West Germany	Pu 02/U02	1,200 MWe	1980

Source: Assembly of Western European Union, Document 686, *Second Generation Nuclear Reactors.* Report by Mr. Lenzer, Paris 1975, derived from Appendix I, II, pp. 10, 11.

Two countries have pursued a supply policy for natural uranium. The first was France, through its state agency, CEA, whose expertise since the 1950s has led to important discoveries of economic reserves not only in France but also in Niger, Gabon, and Canada. Several of the chief countries engaged in nuclear industry—West Germany, Italy, Spain, Japan, and the United States—have been associated in the Niger operations. Next to follow France was West Germany, which in 1975 reached an important agreement with Brazil (see Chapter 3). In the medium term, market security will induce European countries to look toward South Africa, Australia, and the developing countries for their supplies, competing with Japan and with the United States, which will soon be a net importer.[29]

Enrichment of uranium is a large-scale, highly capital-intensive industrial process. The United States, with three plants, is in a position of near monopoly in the Western world. As an alternative, several European countries have turned to the Soviet Union to obtain limited amounts of enriched uranium, and have begun developing enrichment

[29] Ibid., p. 13.

programs based on two different types of technology. A debate arose between those who believed the best way to meet Europe's needs was to play safe and settle for an updated version of the well-established U.S., European, and Soviet technology of gaseous diffusion and those who believed that Europe should take a leap forward by launching centrifuge technology.

Both processes have their drawbacks. The initial plant for gaseous diffusion must have a capacity of treatment of 10,000 tons per year of natural uranium (UF^6), and requires for its operation three or four nuclear power plants with a total capacity of 3,800 MWe. The gas centrifuge is much less power hungry, and can be ordered in small installments. But it is made up of millions of small units that would have to be manufactured on a one-time basis unless new installments were to be built on the site or the parts exported. The critical size of the centrifuge is greater than was initially expected; furthermore the plant may require very large maintenance facilities.

A dual plant using both processes relating additional investments to growing needs, might have been a promising possibility, but political considerations acted against such a measure. Britain, West Germany, and the Netherlands had reached similar stages of development and felt confident of the centrifuge method. In 1970 they decided to form a joint venture, Urenco, building two experimental plants in Britain and in the Netherlands. France, which had refused to join Urenco, assumed leadership in a cooperative gaseous-diffusion venture under the name of Eurodif. The French initiative was rejected by the three members of Urenco and later by Sweden, but was welcomed by Spain, Italy, and Belgium. In 1975 Iran joined the venture, assuming part of the French share. Eurodif works were started in 1974 at Tricastin (France), with production scheduled for 1978.

In view of the rapid increase in demand and the long lead-time, a study group by the name of Coredif comprising the same partners as Eurodif, with a different distribution of shares, is now considering the construction of a 5 million Separative Work Unit (SWU) plant to be added to the 10.8 million SWU plant in Tricastin.[30]

The two Urenco centrifuge plants are expected to provide 2 million SWU by 1985. Under present circumstances in Europe, both processes are needed, and natural uranium can be imported from ore-rich countries of the OECD area. At a later stage, European and especially Community industry will have to become involved in the development of enrichment facilities in producer countries such as

[30] A Separative Work Unit (SWU) is a measure of the work required to separate the fissile isotope U 235 from natural uranium.

Canada, Australia, and South Africa, and will have to import enriched uranium supplies from those countries.[31]

Reprocessing irradiated fuels constitutes both a uranium-recovery operation and a plutonium-producing activity, an essential feature of a nuclear-fuel policy linked at present to light reactors and in the future to fast-breeder reactors. Because of the high costs of reprocessing, Britain, France, and West Germany have joined forces to create a cooperative organization, United Reprocessors Gmbh, with the purpose of increasing the capacities of French and of British plants up to 1980 and of building a new plant in West Germany. The latter, barring siting obstacles, should become operational in 1983–84. A shortage of reprocessing facilities may meanwhile arise between 1980 and 1984.[32]

Industrial Structure. European governments, deeply involved in nuclear energy, are anxious to maximize the contribution of their national industries in establishing the programs of power stations and of the fuel cycle. The smaller countries are forced to invite foreign companies to participate in public investment. The larger countries systematically abstain from doing so.

They tend to promote mergers so as to back one national company. As a result, markets in the large countries are compartmentalized, with a monopolistic or oligopolistic structure. Closed to European competitors, they are open to the U.S. manufacturers through licensing arrangements. Orders are often erratic because of the discontinuity of programs. Fluctuations in demand are amplified by the growing size of reactors—from 1,000 to 1,300 MWe—and of turbine generators.

In view of the high capital costs, of the large technical and financial risks, and of the comparatively small number of orders per year, only very large firms can be competitive as prime contractors for the design and construction of nuclear plants and can coordinate the supply of components from various subcontractors.[33]

Out of 300 reactors in operation or on order at the beginning of 1974, only seventeen were based on other than U.S. technology. The U.S. manufacturers had exported forty-four reactors and had built thirty-eight abroad under licensing agreements.[34] Besides Westing-

---

[31] EEC, *Nuclear Fuel Supply Policy*, pp. 10, 14.

[32] Ibid., p. 23.

[33] Surrey, "Future Growth of Nuclear Power," pp. 108-09.

[34] Commission of the European Communities, *Report on the Situation and Prospects of the Heavy Electro-Mechanical and Nuclear Equipment Industries* (Sec) 75, 2770 (Brussels, 1975), para. 5.1, p. 29 (French edition).

house and General Electric, and two other smaller U.S. competitors on the market, about ten European companies are offering their services either as prime contractors of proven reactors or as manufacturers of pressure vessels. Comparing the European and U.S. industrial structures, Europe has twice as many companies for power plants and three times as many for pressurized vessels in a market half as large.[35] Obviously European companies cannot achieve economies of scale or sustain levels of research, engineering, and marketing efforts comparable to those of their chief U.S. counterparts. It is generally assumed that three or four nuclear plant orders a year are necessary to stay in business.[36]

The two largest producers of nuclear reactors in Europe, on the basis of present national programs, are West Germany and France. KWU is the European giant. It benefits from the worldwide reputation of Siemens, and is collecting orders from overseas markets such as Iran, Brazil, and South Africa. Siemens has mastered an independent technology for PWRs. In France, Framatome is now the sole prime contractor for proven reactors. It has become a semi-public company, with the French nuclear agency CEA acquiring from Westinghouse a sizable share in the capital stock, and with Creusot-Loire, a company of the Belgian Empain group, due to be the majority shareholder when the Westinghouse license expires. Framatome is actively prospecting foreign markets and has collected orders in Belgium and South Africa.

Brown Boveri and Company, originally based in Switzerland, has expanded considerably in Germany and now controls companies in France and Italy which manufacture major components. It is the only European multinational group engaged in the nuclear industry in several European countries. The Swedish company Asea-Atom has meanwhile developed an original and efficient BWR technology. In Britain, the National Nuclear Corporation, in which private industry is associated with the Atomic Energy Authority, has limited prospects because of the experimental nature of the type of reactor chosen and the small dimension of the nuclear program.

Smaller European states have been forced to enter into the orbit of the larger ones, or to rely on intercompany ties. France and Belgium and France and Italy are closely related at the public-utility and at the technical level, while German industry is influential in Switzerland, Austria, and the Netherlands. The two major American multi-

[35] Ibid., para. 6.1, p. 38 (French edition).

[36] Surrey, "Future Growth of Nuclear Power," p. 122.

nationals have licensees in most of the European countries committed to nuclear power, with the exception of Britain.

In the related field of heavy electrical and mechanical equipment, there is a proliferation of dependent-component suppliers, resulting in over-capacity (in the case of power transformers) and in manufacturing gaps (in the case of very large steel ingots for high-power generators which must be imported from the United States or Japan). According to a recent study, the profitability of this industrial branch is low.[37]

Taken as a whole, the structure of the European nuclear industry is uneconomical. There is a fragmentation of means and a waste of competence as a result both of the smallness of markets and of the anarchy in interstate or intercompany agreements with different partners for different phases of the manufacturing process. The consequences of the situation are overinvestment and higher costs and prices, which in turn hamper the expansion of this high-technology industry in its own growth and in the markets of the developing countries where U.S. multinationals are in a strong position. There is obviously a case for streamlining and concentration in Europe through crossnational mergers. The prospects for such a reform are virtually nonexistent, in view of the nationalistic attitudes of European governments.

In the 1960s Western Europe lost the battle of the natural-uranium reactors. This led to the dependence on American LWR technology and consequently on the American supply of enriched uranium. Subsequently, Europeans had a chance to achieve a breakthrough with HTR technology but have probably lost it, despite a lingering possibility of success in West Germany. Europeans are now relying on the breeder reactor in a dispersal of efforts which may have the initial advantages of allowing research strategy on a broad front and the choice between several reactor designs and prototypes. In the words of A. J. Surrey, "If the potential benefits of competitive development are to be achieved, it will require a readiness to drop unrealistic ambitions for technological leadership and to enter into licensing arrangements." [38]

Interstate Relations. Much has already been said about interstate relations. European governments have been trying to reconcile na-

[37] Commission of the European Communities, *Report on Electro-Mechanical and Nuclear Equipment*, pp. 15, 19, 44.

[38] Surrey, "Future Growth of Nuclear Power," p. 224.

tional industrial policies with loose interstate or intercompany associations, rejecting both the strong international organization and the cross-border merger.

Euratom, never an accepted leader in European nuclear research and development, survived its quarter-century crisis to embark in 1975 in joint research on nuclear-fusion power, for which the European Community will spend about $300 million.

The European Economic Community has prepared policy reports on the electricity sector and on nuclear fuel supply. These reports, however, have not led to clearcut decisions on the issue of interstate relations. According to the reports, conventional and nuclear electricity will receive approximately two-thirds of the investment to be made in the energy field. The Community average rate of self-financing in the electricity sector reached 50 to 60 percent in 1973. A coordinated financial policy would facilitate the raising of a large quantity of long-term funds. To this end, the European Commission suggests an increased use of the European Investment Bank and of direct borrowing by the EEC.[39] Community loans would be better received if lenders were assured that duplication of investments would be avoided.

A concern common to all countries is the establishment of standard safeguards and controls over the circulation of nuclear material from nationally controlled reprocessing facilities. The spread of nuclear energy poses the danger of proliferation of nuclear weapons capabilities. The United States has proposed the establishment of nuclear-fuel centers on a multinational regional basis.[40] This proposal will probably be studied by Western European countries under the auspices of an international organization such as the EEC or the IEA.

In addition, at the initiative of the United States, seven nuclear powers—the United States, the Soviet Union, Canada, Japan, West Germany, Britain, and France—have reached a secret agreement to synchronize national policies concerning the sale of reactors, enriched uranium, and nuclear technology to nonnuclear countries so as to prevent the dissemination of nuclear weapons. The seven nations are committed not to take unfair advantage of each other, and to consult about sales of equipment.[41]

---

[39] EEC, *Guidelines for the Electricity Sector in the Community* (Brussels, 1974), Com (74), 1970 Final, pp. 40-41.

[40] Address by Secretary of State Henry Kissinger before the United Nations General Assembly, September 22, 1975.

[41] David Howard, "Seven Nations Reach Accord on A Exports," *International Herald Tribune*, January 30, 1976.

Conclusion

The purpose of this chapter was to examine the way in which the countries of Western Europe were utilizing their domestic resources, in the form of either energy conservation or of energy substitution.

Efforts to improve the demand/resource relationship have met with constraints that hamper the adoption of one option or of a combination of options. Short-term decisions on consumption of energy and investment in energy have a bearing on options open in the medium term. A constraint common to all energy options—oil and nuclear energy today, innovated coal and solar energy tomorrow—is that they are largely capital-intensive. Therefore costs are strongly dependent on interest rates. As already mentioned, the inflation record of the European countries is not encouraging; the situation is reflected in the rising costs of oil exploration, production, and transportation in the North Sea and of construction of conventional nuclear power stations.

Under these circumstances, an attempt to increase energy efficiency over the transformation chain and at the consumption level is essential. It is perhaps premature to assess achievements in energy conservation since 1973. However, the International Energy Agency has compiled a report rating the performances of the member states. So far these performances are not impressive.[42] The public at large is not required to make any substantial sacrifice in energy consumption, and it is not even aware that Western Europe is in a most precarious situation. Therefore, it is not reacting to the predicament. The quest for energy substitution is being pursued far more on a national than on a collective basis, but with some lack of determination.

In the field of oil and gas, the major but unavoidable economic constraint is uncertainty about the role of domestic resources. There are also internal causes of tension between the haves and the have-nots, the oil- and gas-rich countries and the others. Possibly in Britain and certainly in Norway production over the next decade is expected to be below the capacity of proven reserves. No coherent joint-development program has been adopted for the North Sea.

In the nuclear field, one of the major constraints is technological: the feasibility at the industrial level of the breeder reactor—the most attractive type from the viewpoint of overall energy efficiency and uranium supply—is not yet proven. Another constraint is psychological: the lack of acceptance of nuclear energy by the population.

[42] OECD, *Energy Conservation in the International Energy Agency, 1976 Review*, Paris, 1973, pp. 7-8.

Lastly, the political issue of national prestige reinforces the fragmentation of Western Europe into small markets, to the detriment of industrial efficiency and profitability. Cooperation in nuclear energy has been a patchwork affair and tardily developed.

No pooling of financial resources, either for oil and gas or for nuclear energy, can be effected in the absence of an economic and monetary union.

It is not surprising, in view of all these problems, that the targets for domestic primary energy production and consumption should require revision downwards. At the end of 1975 the EEC Commission reviewed the objectives for 1985 that had been adopted in December 1974. At the earlier date, it was expected that the Community could reduce its level of external dependence from 63 to 40 percent between 1973 and 1985 (EEC estimates were close to those of the OECD experts quoted in Chapter 1, Table 1). The review of 1975 led the commission to the conclusion that the level of external dependence would not fall below 50 percent in 1985.[43]

These bleak prospects postpone for several decades the emergence of a relatively energy-independent Europe. They should be considered in the context of a worldwide energy geostrategy.

[43] EEC, *Realization of the 1985 Objectives of the Community Energy Policy* (Brussels, 1976), Com (76), p. 11.

5

EXTERNAL RESOURCES AND ENERGY GEOSTRATEGY

Actors and Conflicts

In January 1975 Secretary of State Henry Kissinger said in an interview with *Business Week*: "You could never implement the energy policy as a purely economic matter. It has been a foreign policy matter since the beginning."

All countries are affected by the rise in the price of energy relative to the price of manufactures and by uncertainty about the future relationship between energy supply and demand. Countries tend to group together or to oppose each other on the basis of a variety of factors—relative power, degree of industrial development, dependence on external sources of primary energy, and reliance on proven or potential reserves of fossil fuels. In addition, political and ideological motivations are superimposed on considerations of energy security.

Broadly speaking, in this maze of associations and rivalries, several groupings of countries and two distinct situations of conflict can be distinguished.

The United States and the Soviet Union stand apart in their capacity as superpowers, each possessing a full-fledged defense system and a clientele of allies and associated states. Over a period of time, the ambiguous interrelations of these two superpowers vacillate between tension and détente. The energy factor now plays a major role in their worldwide strategy.

The industrialized countries linked militarily and/or economically to the United States fall into two associations, the OECD, of which the United States is a member, and the EEC, which comprises only nine of the countries of Western Europe. In both associations, relations with the United States, which is eager to promote a joint strategy for the industrialized oil-consuming countries, play a major part.

The oil-producing countries also fall into two associations, the broader OPEC (see Chapter 2), and the narrower OAPEC, linking the Arab oil-producing countries. OPEC member countries can be considered as moving from the developing to the developed stage of economic growth. Paradoxically, such is not the case for most of the oil-consuming developing countries, which have tended lately to close ranks with OPEC member countries.

This realignment of national interests centers on two interrelated conflict situations, one local, the other global. The local Arab-Israeli conflict, developing over the decades since the creation of the state of Israel, happens to be located close to the Persian Gulf area around which lie more than half of the world's proven reserves of crude oil. The global conflict can be considered as the final phase of the decolonization process. In a resolution of the UN General Assembly in 1966 the international community recognized "the indefeasible right of the States to exert permanently their sovereignty over their natural resources in the interest of their development."[1] The oil-producing countries were the first to fulfill that goal in terms of ownership, rent, and price-maximization, with the developing countries that produce and export other commodities following suit. This global conflict is therefore concerned with the control and pricing of natural resources.

Western Europe, whose bases of energy and raw material have eroded over the last decades, appears to be the loser in both the local and the global conflicts. An evaluation of the European position in general and more particularly that of the larger Western European countries must first consider the energy resources and policies and the major political goals of the two superpowers, the United States and the Soviet Union. Second, it must deal with the situation and prospects in the Middle East, beginning with the October 1973 war. Finally, as a preface to an analysis of the global struggle for natural resources, it must analyze the complex relationship among the Western industrialized countries under U.S. guidance.

The United States

Needs and Resources. Endowed with very large reserves of fossil fuels in proportion to the size of its territory, the United States, with 6 percent of the world population, accounts for 30 percent of total world energy consumption. U.S. oil policy in the 1950s tended to protect the domestic market and to foster the expansion of U.S.-based

[1] United Nations, *General Assembly Resolution*, November 1966.

Table 12

U.S. ENERGY NEEDS AND RESOURCES AS PERCENTAGES OF WORLD TOTALS, 1972

	Proven Reserves	Production	Consumption
Oil	5.6	20	30
Natural Gas	14.0	53	55
Coal	20.0	25	23

Sources: Oil—*BP Statistical Review of the World Oil Industry*, 1972; natural gas—*UN Statistical Yearbook*, 1973; coal—OECD, *Energy Prospects to 1985*, 1974, vol. 1, derived from Table 2-2, p. 47.

multinational companies on the world market. Domestic oil prices were kept high through import quotas and price regulation. Drilling costs were written off as current expenses. A depletion allowance was set up, deductible from taxable income. Tax credits for payments to foreign governments were offset against American corporate income tax liabilities.

By contrast, gas prices were kept low for household consumption. In the late 1960s severe restrictions on strip mining and on the use of sulfur coal kept coal production down. The action of environmentalists stopped offshore drilling along the east and west coasts and delayed by four years the building of the Alaska pipeline. Difficulties with the development of technology and opposition on grounds of safety contributed to slowing down the expansion of nuclear power. This combination of policy regulations and environmental constraints, coupled with a high rate of growth in the economy, led to a rapid increase in oil imports in the early 1970s.[2] These imports accounted for 31 percent of oil consumption in 1972 and for 38 percent in 1974, and respectively for 14.4 and 17 percent of total primary energy requirements.[3] The forecast of import requirements for 1976 is estimated at 42 percent of oil demand.[4]

Table 12 presents the energy needs and resources of the United States as a percentage of the world total in 1972. Coal reserves are high in proportion both to world reserves and to the proven reserves of all fuels in the United States (88 percent).[5] The table does not allow

[2] James W. McKie, "The United States," *Daedalus*, Fall 1975, pp. 74, 76.

[3] *BP Statistical Review of the World Oil Industry*, 1972, 1974, pp. 8, 10.

[4] "Large Oil Imports Expected," *The Petroleum Economist*, July 1976, p. 251.

[5] OECD, *Energy Prospects*, vol. 2, p. 149.

for the unestimated but certainly considerable reserves of offshore oil, nor the oil potentially available in shale, believed to be forty times greater than proven oil reserves.[6] Because of the variety and volume of its proven and potential reserves of energy, the United States can choose among several domestic policy options at different price levels and over different periods of time. Depending on the ability of the government to overcome some of the environmental constraints, major efforts over the next decade could be directed primarily either to West Coast low sulfur coal or to offshore oil, or to nuclear energy. Shale oil will not be a sizable contributor to the energy supply before the 1990s.

Postponement of Energy-Policy Options. No clearcut choice of alternatives has been made since the beginning of the oil crisis. Late in 1973 President Nixon launched "Project Independence," with the announced goal of achieving self-sufficiency by 1980 by expanding domestic oil production and developing new energy sources through higher prices. Price controls over the production of new oil wells were removed, but in spite of the price incentive, production continued to fall, as a result of the slow but steady depletion of reserves.[7] At the same time, the wellhead price control of natural gas was continued though the reserve/production ratio had fallen from twenty-to-one in 1962 to eleven-to-one in 1972.[8]

In January 1975 President Ford changed tactics and outlined an energy program aimed at reducing crude-oil imports by introducing an import duty and by lifting price controls on all domestic production of crude oil. Congress opposed decontrol because it meant a rise in consumer prices. After a long struggle, the President signed a controversial act in December 1975 reimposing price controls on "old" and "new" crude oil for forty months.[9] The end result was to reduce slightly in 1976 the price of petroleum products in the United States. Under the circumstances, oil imports are not expected to be seriously curtailed in the near future. The Exxon Company takes the pessimistic view that imports will grow from 6.4 million barrels per day in 1975, in other words, 40 percent of liquid demand, to 10 million barrels per day by 1980, or 50 percent of liquid demand.[10] Only in 1976 after

[6] Philip W. Quigg, "New Energy Sources," in *Economics of Energy*, Leslie E. Grayson, ed. (Princeton, N.J.: The Darwin Press, 1975), p. 437.

[7] *The Morgan Guaranty Survey*, June 1975, p. 8.

[8] OECD, *Energy Prospects*, vol. 2, p. 119.

[9] "No Energy Policy in USA," August 1975, p. 291, and "USA Prolongs Price Controls," January 1976, p. 10 in *The Petroleum Economist*.

[10] "World Energy Outlook," Exxon Press Briefing, New York, December 9, 1975.

several years of interruption, did the government proceed to sell offshore leases in the Gulf of Mexico, on the Atlantic Coast, and in the Gulf of Alaska.[11]

According to a 1975 unpublished report by the International Energy Agency, the oil conservation program is "neither strong nor comprehensive" in the very country in which per capita energy consumption has always been the highest.[12]

Conclusions were drawn from the above by two authorities in the field of American energy policy. The first is from James W. McKie: "The long run allocation of industrial energy requirements among coal, oil, and nuclear power is still in debate." [13] The second is from Mason Willrich: "The government's persistent inability to make political choices among alternative *means* may also reflect doubt and ambivalence about energy independence as a national goal." [14]

Energy Dependence and Superpower Constraints. The present lack of political consensus has very serious implications in the international setting, where the United States is subject to the constraints of a superpower.

The degree of energy dependence affects the relations of the United States both with its suppliers and with its allies. The predicted increase in the volume of U.S. oil imports will come from the Middle East, since Canada intends to reduce its supplies to the United States and Venezuela plans to stabilize the level of its production. This increase will strengthen the bargaining position of OPEC as a whole at a time when Western Europe and Japan are striving to maintain their imports from the Middle East at a plateau.

The economic viability of Western Europe and Japan, and consequently the political basis of the alliances linking these two areas to the United States, are presently dependent on the continuous flow of oil from the Middle East and Africa. Again in the words of Mason Willrich, "A growing vulnerability to interruption in its oil supplies is dangerously incompatible with America's specific responsibility as guarantor of the security of the NATO countries, Japan and Israel." [15]

At the same time that the United States is committed to maintaining the existence of the state of Israel and therefore to containing

---

[11] "The Search Moves Offshore," *The Petroleum Economist*, May 1976, p. 165.

[12] James Goldsborough, "Energy Agency Rates U.S. Poorest in Oil Conservation," *International Herald Tribune*, November 27, 1975.

[13] McKie, "The United States," p. 76.

[14] Mason Willrich, "Energy Independence for America," *International Affairs*, January 1975, p. 53.

[15] Ibid., p. 66.

the expansion of the Soviet Union in the Middle East, it must develop close and stable relations with the key oil-producing countries, mainly Saudi Arabia and Iran. The United States is also concerned with the military balance in the area. It is supplying arms to Israel, but not to the point of providing this country with an indisputable superiority over its Arab neighbors. It is selling a considerable amount of modern weaponry to Saudi Arabia and to Iran (not to mention sales to smaller Arab states), in order to deter the Soviet Union from intervention in the Persian Gulf. It maintains powerful naval forces in the Mediterranean and in the Indian Ocean, to protect oil shipments from the Middle East and North Africa to Europe, and from the Middle East to Japan.

In regard to international oil security, the United States occupies the pivotal position in the world.[16] Its complex network of superpower obligations puts some strain on its relations with its European associates in energy-related matters. At the same time the plight of Western Europe has given the United States an opportunity to assert its leadership in the Atlantic Community. Prior to exploring the Euro-American relationship, some light must be shed on the energy policy of the Soviet Union and on Euro-Soviet energy linkage.

The Soviet Union

The Soviet Union has considerable reserves of fossil fuels, two or three times as large as those of the United States. A geological survey of these resources is far from complete, and important discoveries have recently been made in the remote territories of northern Siberia. The bulk of the new reserves is located at a great distance from manufacturing centers and in inhospitable surroundings, bearing on costs of both production and transportation.

Over the last two decades, Soviet production and consumption have shifted from coal as the dominant source of energy to oil and gas. The ratio of crude-oil reserves to current annual output is ten-to-one.[17] The ratio of gas reserves to production is much higher. The growth in production, though sizable, has fallen below targets, because of technological difficulties related to soil and climate, and also because of a shortage of modern drilling and transportation equipment.

Table 13 presents the energy needs and resources of the Soviet Union in percentage of the world total in 1972. Net exports of primary

[16] Ibid., p. 59.

[17] Yager and Steinberg, *Energy and U.S. Foreign Policy*, p. 194.

Table 13

U.S.S.R. ENERGY NEEDS AND RESOURCES AS PERCENTAGES OF WORLD TOTALS, 1972

	Proven Reserves	Production	Consumption
Oil	14.7 [a]	15.0 [a]	12.7 [a]
Natural gas	34.4 [b]	18.4 [b]	18.4 [c]
Coal	62.0 [b]	21.0 [b]	20.0 [c]

Sources: [a] *BP Statistical Review of the Oil Industry*, 1972; [b] *UN Statistical Yearbook*, 1973; [c] U.N. Statistical Office, *World Energy Supplies.*

energy amount roughly to 10 percent of total production. The pattern of trade is directed by a combination of political and geographical considerations, notwithstanding the opportunity of earning foreign currencies.

Since the early 1960s Eastern Europe has become a net importer of primary energy. It is linked by pipeline to the Volga Ural oilfields and to the Ukrainian gasfields. For years the Soviet Union remained the sole supplier of hydrocarbons to its Comecon partners. In 1972 deliveries of crude oil from the Soviet Union, on the basis of long-term contracts, accounted for about two-thirds of Eastern Europe's consumption. At present, because of the growing energy requirements of both the Soviet Union and its partners, the latter have been authorized to purchase crude oil directly from Middle East producers.

The Soviet Union has been a regular but secondary supplier of oil, mainly to Italy, but also to Sweden, West Germany, and France. However, oil shipments have declined in relation to European consumption, from 9 percent in 1960 to 7.2 percent in 1972.[18] By contrast, gas shipments are on the rise, with Austria the first purchaser and West Germany at present the largest. Contracts have been signed with Italy and France, and Sweden and Switzerland may follow suit. Gas deliveries are paid for partly in kind, in the form of large-diameter pipes to extend the Soviet pipeline network from the west Siberian fields to Eastern and Western Europe. A triangular pact was signed in 1975 between Iran, which agreed to increase its regular sales of gas to the Soviet Union, and the Soviet Union, which agreed to increase by the same quantity its sales to West Germany, France, and Italy. By 1985, Western European imports of natural gas from the Soviet Union

[18] *BP Statistical Review of the World Oil Industry*, 1972, pp. 8, 10.

are expected to represent approximately 5 percent of the area's total gas consumption.[19]

Present and prospective gas transactions are favorable to both parties. The Soviets obtain equipment, and Western Europeans diversify their energy sources. From the levels quoted above for oil and gas supplies, it seems unlikely that Western Europe will become dangerously dependent on the Soviet Union for energy.

As in the case of the United States, the energy policy of the Soviet Union is subject to superpower constraints. Since World War II, the Middle East has been a contested zone of influence between the Soviet Union and the United States. Iraq and Syria belong to the traditional clientele of the Soviet Union, while Egypt, for a number of years in the Soviet orbit, has changed camps. These three countries have purchased sophisticated arms from the Soviet Union, which the Egyptians used during the Yom Kippur war. Meanwhile, the Soviet Union is asserting its naval presence through the important buildup of its Mediterranean and Indian Ocean fleets, which have bases in Somaliland and South Yemen, and have gained flexibility from the reopening of the Suez Canal. In short, the Middle East remains an area of superpower confrontation despite détente. If, as Mason Willrich pointed out, conservative governments in a few Gulf states were to be replaced by radical governments, and if the policy of détente were to be reversed, the Soviet Union might be tempted to exploit U.S. oil dependence (and also that of Western Europe and of Japan).[20]

The Middle East

The conflict in the Middle East, from the point of view of Israel and its Arab neighbors, culminated in the October 1973 war. From the point of view of relations between Arab oil producers and Western oil consumers, on the other hand, the conflict culminated in an embargo against selected industrialized countries which lasted until mid-1974. In the period between war and embargo, the Common Market countries strove to engage in a Euro-Arab dialogue, while the United States concentrated its efforts on achieving partial settlements between the former belligerents. But in spite of the September 1975 agreement, in which the United States became directly involved, the Middle East situation is far from being stabilized, and is shrouded in uncertainties, in the short as well as in the long term.

[19] OECD, *Energy Prospects*, vol. 1, derived from Tables 2-5 and 2-6, pp. 50-51.
[20] Willrich, "Energy Independence," p. 59.

The October War. The United States assumed two major responsibilities during the October war: first, to save Israel through an airlift, and second, to impose a cease-fire in time to prevent a military invasion of Egypt. The airlift was launched without previous consultation with Western Europe, and was implemented without its assistance except for Portugal, which did not object to the use of Azores bases for this purpose. Britain, on the other hand, went so far as to declare an embargo of arms shipments to all belligerents. In the same way, the United States did not consult its NATO partners when it launched a worldwide nuclear alert to prevent military intervention by the Soviet Union in favor of the Arabs. Finally, the cease-fire was arranged by the United States without European support. These successive U.S. interventions were the actions of a superpower aware of both the local and the global dimensions of a still-limited conflict.

The Western European countries (and Japan) viewed the crisis in terms of economic security. They were unable to influence military developments in the Middle East and were without any prospect of an immediate substitute from the United States for Arab oil. In a kind of panic they refused to facilitate the flow of U.S. arms to Israel and instead took sides with the Arabs.[21]

Thus there was a profound discrepancy between American actions and the way the actions were perceived by Europeans. The United States resented the refusal of its allies to cooperate on the airlift, and the Europeans resented the failure of the United States to consult them on its Middle East strategy. Atlantic Alliance procedures proved inadequate to handle an extra-alliance conflict.

The Oil Weapon. The Arabs decided to use economic pressure on the Europeans to force a change in the U.S. attitude toward Israel: such was the purpose of the oil weapon, which consisted of restricting oil supplies while the price was raised. While all OPEC member states, Arab and non-Arab, applied the price rise, the oil cut-back and embargo were debated in the Organization of Arab Petroleum Exporting Countries (OAPEC), of which Egypt and Syria were members. OAPEC placed an embargo on oil supplies to the United States, Portugal, the Netherlands, and Denmark, together with successive cutbacks in production reaching a maximum of 25 percent by the end of 1973. France, Britain, and Spain, in view of their support of the Arab cause, were declared friendly countries and not subject to restrictive measures.

21 Klaus Knorr, "The Limits of Military and Economic Power," *Daedalus*, Fall 1975, p. 230.

The reason for the boycott of deliveries to the United States and Portugal was obvious. The Netherlands, on the other hand, was penalized because it had blocked a proposal in the United Nations Security Council to let France and Britain act as spokesmen of the EEC countries during the debate on the Arab-Israeli conflict. France and Britain strongly backed the declarations adopted by the EEC Council of Ministers on November 6, 1973, calling for Israel to implement fully UN Security Council Resolution No. 242, an interpretation similar to the Egyptian position on state borders and occupied areas. West Germany was forced to align itself with France and Britain because much of its supply of oil and petroleum products passed through Rotterdam, then under Arab embargo. In March 1974 West Germany and Italy were added to the OAPEC list of friendly nations. The embargo on the United States was finally lifted in March 1974, and on the Netherlands in July 1974.

The embargo was on the whole ineffective. The governments of OECD countries delegated to the oil companies the responsibility for allocating oil supplies so as to equalize the burden. The European countries had oil for about ninety days' consumption in storage, and for about thirty more in shipment. At no time was there a real shortage of petroleum on the European market.[22] Europeans yielded to the pressure of oil-producing countries exploiting a relationship of asymmetrical economic interdependence. No single commodity was as indispensable as petroleum to the European countries, and no economic reprisal could be a threat to OPEC.[23]

The selective embargo produced divisions and tensions within the European Community. France and Britain opposed oil-sharing within the EEC after the Netherlands had been declared a hostile state by OAPEC. Intended as a friendly gesture towards the Arabs, this attitude was of course damaging to European solidarity. Several of the European countries tried to come to terms with individual oil producers by making bilateral trade deals. On the whole the behavior of member states when the oil weapon was being used proved that the Community tended, in Klaus Knorr's words, "to fall apart under adversity."[24]

The Euro-Arab Dialogue. In spite of disagreement on the scope and content of a future common energy policy, the nine EEC member coun-

22 Prodi and Clô, "Europe," pp. 98, 101.

23 Knorr, "Limits of Military and Economic Power," pp. 232-33.

24 Ibid., p. 239.

tries attempted, under French initiative, to engage in a so-called Euro-Arab dialogue. Responsibility for the talks was entrusted on the European side not to the Brussels Commission but to a special committee dealing with foreign policy cooperation, the Davignon Committee, and on the Arab side not to OAPEC but to the Arab League. The United States, regarding this initiative as a complicating factor in an area where they assumed major diplomatic responsibilities, succeeded in removing the subject of oil from the agenda of the preliminary meetings, which took place in Cairo in June and in Paris in July 1974. This omission left industrial, financial, and cultural cooperation as main items for discussion. The political issue of representation of the Palestinian Liberation Organization (PLO) troubled the dialogue. Military issues played no part, since Western Europe offered no alternative in the security field. The dialogue seemed unlikely to become an international force molding events.[25]

The Egypt-Israel Settlement. Since the October war, the United States, aside from its continuous and complex relationship with the Soviet Union, has focused its diplomatic action on two related fronts: the Middle East and the international economic implications of the oil crisis (discussed below). The first result of Secretary of State Henry Kissinger's "shuttle diplomacy" was to arrange for a preliminary agreement in January 1974 between Israel and Egypt on the withdrawal of forces. After a number of missions in the field, Secretary Kissinger was able to sponsor in September 1975 an agreement of wider scope between the two countries. The withdrawal of Israeli troops from the Sinai Peninsula was completed, and the state of belligerency between the two countries came to an end. Israel was authorized by the agreement to use the Suez Canal, reopened in June 1975, for nonmilitary shipments. Israeli and Egyptian armies were to be separated by a buffer zone under control of UN forces, with U.S. civilian experts operating tactical early warning stations. Thus the United States became actually the guarantor of peace between the two countries.

The position of the United States in the Middle East has undergone dramatic improvement since 1973. Relations of mutual good faith have been established with Egypt. The United States is also on improved terms with Iran, whose general interests converge with those of the United States, particularly since an agreement between

[25] Vittorio Barattieri and Judith D. Trunzo, *The Euro-Arab Dialogue*, forthcoming article.

Iran and Iraq over the Kurd rebellion, concluded in March 1975, has made Iraq less dependent upon the Soviet Union. Nonetheless, global settlement of the Arab-Israeli conflict requires an agreement with Syria, Jordan, Lebanon, and the PLO, and therefore entails the cooperation of the Soviet Union within the framework of the UN Geneva Conference.[26]

The record of U.S. diplomacy in the Middle East shows the paramount importance of this area in the geostrategy of the United States. Because of European, Japanese, and (to a lesser extent for the time being) U.S. dependence on Middle East oil, the area is considered vital. Such is the significance of statements made by President Ford, Secretary of State Kissinger, and former Secretary of Defense Schlesinger that U.S. military intervention might be necessary if the oil-producing countries threatened "some actual strangulation" of the Western economies.[27]

Uncertain Prospects. The short-term prospects of Arab-Israeli conflict are still disquieting. Lebanon was torn by civil war. Palestinian territorial claims are far from being settled. The timing of a general settlement remains unpredictable.

In the medium term, future prospects in the Persian Gulf area must be carefully considered. The key country is Saudi Arabia, with its enormous oil reserves, low-cost production capacity, and surplus revenues. Its oil minister remarked that Saudi Arabia, with an authorized production level of 8.5 million barrels per day and requirements of only 3.5 million barrels per day, was overproducing.[28] As stressed in Chapter 2, one reason OPEC is a strong cartel is that Saudi Arabia is in a dominant position and can to a large extent adjust production to demand so as to maintain an agreed level of prices. What if a political upheaval took place in this country? Basically Arab countries do not want to fall under Soviet domination, at least as long as the present conservative governments remain in power. But an oversupply of sophisticated arms could present a temptation to ambitious radical officers. The fantastic arms buildup which is taking place in the Gulf area is a powerful destabilizing factor. How would the Soviet Union react to a situation of instability? Would it try to exploit a local crisis, and thereby endanger once more European

[26] Jacques Vernant, "Politique et Diplomatie," *Défense Nationale*, June 1975, p. 130; November 1975, pp. 89-91.

[27] Ibid., p. 132; Willrich, "Energy Independence," p. 58.

[28] Ahmed Zaki Yamani, "A New Era of Stability in Energy Economics," *Journal of Energy and Development*, Autumn 1975, p. 5.

economic security, or would it continue to bank on détente? [29] This is an open question, mentioned above in Chapter 5. Whatever the outcome, European attitudes would be irrelevant in a revolutionary situation in the Middle East, since there are only two major defense systems. Europe contributes nonetheless to local instability first by its arms sales (although at a much lower level than those of the United States), and second by the sale of nuclear power plants—Iran, Egypt, Saudi Arabia, and Libya are on the shopping list—which open vast possibilities for nuclear proliferation.[30]

However, the effects of the Middle East war in 1973 and the possibility of political upheavals in this area in the future should not lead to the conclusion that the price of oil, and in a broader context the price of energy, are not influenced by market forces. In 1973 a boom in demand took place almost simultaneously in all the industrialized countries. The volatility of the market helped the OPEC countries acting as an oligopoly, to increase considerably their producers' rent. The 1974–75 recession forced the industrialized countries to curtail their imports somewhat, and the producer countries to freeze the price of oil at a time of strong inflation—in effect, to reduce the price in real terms. As industrial production caught up in 1976, especially in the three leading countries, the United States, Japan, and West Germany, oil prices were almost bound to surge and an actual increase of 5 and 10 percent did actually take place at the end of the year. The only influence the oil-consuming countries can exert on the market is through energy conservation and energy substitution, the effects of which are felt in the medium or long term.

These remarks are intended to put political and military factors in the Middle East in perspective in relation to the energy market. It is up to the industrialized countries to take stock of this interrelationship.

Relations between the Industrialized Countries

Interaction between the industrialized countries takes place continuously in the European Community, and in the OECD and its operational agency, the IEA. The United States is the obvious and almost

[29] Uwe Nerlich, "The Superpowers in the Middle East; Stakes, Incentives and Obstacles for Coordinating West European Policies," a paper prepared at the third meeting of the Western European Study Group on the Middle East and the Crisis in Relations among Industrial States, Atlantic Institute for International Affairs, 1975.

[30] Hanns Maull, "Oil and Influence: The Oil Weapon Examined," in *Adelphi Papers*, no. 117 (London: The International Institute for Strategic Studies, 1975), p. 23.

unchallenged leader at the OECD level, with U.S. influence and leadership in the background of most of the energy policy debates at the EEC level.

The aftermath of the Yom Kippur war left the European countries in a state of disarray and division. Oil-rich Britain showed little willingness to share its potential resources with its Community partners. The Netherlands, as the seat of one of the major oil companies, opposed any initiative of the Brussels Commission to control the functioning of the energy market along the lines of French practice. In addition to the division between oil-rich and oil-poor countries, there was a division between countries such as West Germany and the Netherlands that had strong currencies and were able to absorb the economic consequences of higher oil prices, and countries such as Italy, Britain, and more recently France, whose currencies were threatened by a combination of domestic troubles and increased cost of oil imports. Finally, there were diverging views regarding relations with the United States, as France opposed and its partners accepted more or less wholeheartedly U.S. leadership in the management of the oil crisis.

The Europeans took no joint initiative during the fall of 1973, except to support Arab views in the Middle East conflict. It was Secretary of State Henry Kissinger who, in his speech at the Pilgrim's Society in London in December 1973, advised consumer countries to oppose the OPEC cartel and try to force it to lower oil prices. A conference was convened to this effect in Washington in February 1974. The French government, after some hesitation, decided that its foreign minister, Michel Jobert, would attend, on condition that EEC member states would take a common stand. Jobert argued that no permanent consumer organization should be set up, that OECD was the proper framework for Atlantic technical cooperation, and that a world energy conference should be convened. These proposals ran counter to Secretary Kissinger's proposal for a scheme for emergency sharing of oil and the creation of an Energy Coordination Group, with the task of creating a consumer organization.

The U.S. view prevailed, however, as a result of which IEA was launched a few months later as a specialized agency of the OECD. The United States hoped to promote a common energy policy for the industrialized countries, but France considered such a move as leading to a dangerous "confrontation" with the oil producers. However, Michel Jobert's rigid attitude brought the Common Market countries, except France, closer to the United States. France had overplayed her hand.

The French refusal to join the IEA made it all the more difficult to establish a consensus among the industrialized countries. Procedural obstacles were encountered when the IEA and the EEC had to discuss such overlapping items on their agendas as emergency sharing, floor price, market regulation, and joint research and development programs.

As time passed, French and U.S. attitudes regarding relations with the oil-producing countries converged (see below), and consequently the relationship between the two international organizations improved, and their respective roles became more distinct. The IEA was created to fulfill an operational function. The Energy Committee of the OECD (of which France is a member) was reestablished in January 1976 and resumed its advisory function on policy.

Nonetheless, the competence of a divided European Community to deal with energy problems was questioned. It was argued that the EEC framework was challenged by smaller and larger formations, by bilateral agreements on the one hand and by the OECD–IEA compound on the other. France's EEC partners were of the opinion that it was neither possible nor desirable to define an energy policy in the framework of the nine EEC countries without including the United States and Japan.[31] It was also argued that the energy crisis arose almost completely from political and economic conditions extraneous to Europe and that in the crisis Europe was a protagonist of the second rank.[32] According to the view of Miriam Camps there still seems "to be scope for a distinctive energy policy on the part of the Nine, if they choose to take that road." [33] The major EEC countries do not have a common basis for an energy policy. A case in point already mentioned in Chapter 4 is the reluctance of France concerning the fixing of a floor price for imported oil.

Nonetheless, this writer is of the opinion that in a worldwide debate on energy and related issues, the EEC is a useful intermediary for the United States, the non-oil-producing consumer countries, and OPEC, as testified by the recent evolution in producer-consumer relations.

Producer-Consumer Relations

Two parallel schools of thought developed in international circles once the shock of the oil crisis began to recede. The first arose in the Euro-

[31] Jacques Vernant, "Politique et Diplomatie," April 1975, p. 105.

[32] Prodi and Clô, "Europe," p. 91.

[33] Miriam Camps, "'First World' Relationships: The Rule of the OECD," *The Atlantic Papers*, February 1975, p. 33.

pean Community, with France as its strongest advocate. The argument ran that the EEC, as the greatest oil importer, should establish lasting ties with the oil producers. The second school of thought surfaced in the various forums where developing countries met. The basic idea was that the less fortunate developing countries should join forces with OPEC to improve their economic prospects by adding the weight of OPEC's strong bargaining power with the industrialized countries. Conversely OPEC countries saw the opportunity of acting as leaders of the Third World. EEC authorities also, especially the French, were of the opinion that in the long term there were, in the words of Richard Wigg, economic advantages in "creating real purchasing power in the raw-materials-producing developing countries."[34] A complex interplay of diplomatic action and public statements led to the convening in December 1975 of what was initially called the North-South Conference, and later named the Conference on International Economic Cooperation (CIEC). It is in the light of these developments that the main stages of producer-consumer relations since 1974 must be reviewed and the prospects for CIEC assessed.

The present claims of developing countries in regard to the industrialized countries were formulated by President Houari Boumédiènne of Algeria at the UN General Assembly in April 1974 under the title of a "New International Economic Order" (NIEO). A denunciation of the international status quo was also reflected in the Charter of Economic Rights and Duties of States adopted by the UN General Assembly in December 1974. A few months later, at the Conference of Developing Countries on Raw Materials in Dakar in February 1975, the Third World oil-importing countries showed an impressive solidarity with the OPEC countries.[35]

In Geneva in August 1974, Sheikh Yamani, the Saudi-Arabian oil minister, proposed that states belonging to three categories—oil-producing, industrialized, oil-importing developing—discuss NIEO problems, including the consequences of the oil crisis. President Giscard d'Estaing adopted the Saudi list (with Saudi Arabia, Algeria, Iran, and Venezuela as oil-producing nations; the United States, the EEC, and Japan as industrialized; Brazil, India, and Zaire as oil-importing developing) and started diplomatic consultations, which culminated in his meeting with President Ford in Martinique in December 1974.

[34] Richard Wigg, "France and the Raw Materials Question," *World Today*, December 1975, p. 503.

[35] Lincoln Gordon, "The U.S. and the Third World" (Washington, D.C.: Woodrow Wilson International Center for Scholars, June 1975).

President Ford agreed to a tripartite conference, provided the industrialized countries showed solidarity.[36]

In January 1975, the OPEC ministers, following NIEO concepts, asked that discussions should not be limited to oil, but extended to raw materials and development, a suggestion acceptable to the EEC. At the same time, in January 1975, the European Community signed in Lomé (Togo) an agreement with forty-six countries of Africa, the Caribbean, and the Pacific, which met some of the claims of the Third World. An important provision of the agreement guaranteed the stability of receipts of the raw-material-exporting countries. At a preparatory meeting of the North-South Conference in Paris in April 1975, the participants were invited to debate on "energy and related problems of international economic relations," but the United States refused to discuss any problem other than energy, and consequently the meeting was adjourned. Subsequently, in May 1975, at the Commonwealth meeting in Jamaica, Prime Minister Harold Wilson expressed strong support for commodity agreements, thus confirming European sympathy for Third World views.

Confronted with the converging views of OPEC, the non-oil-producing developing countries, and the European Community, Secretary Kissinger reconsidered his position. In a speech in Paris on May 27, 1975, at the ministerial meeting of the IEA, he announced that he was ready to broaden the dialogue between producer and consumer countries to include "all the issues of concern in developing countries."[37] On September 1, 1975, at the Seventh Extraordinary Session of the United Nations General Assembly, Secretary Kissinger asserted that the world economy constitutes a single global system of trade and monetary relations, on which depends the development of all national economies. He specifically favored commodity agreements,[38] (though his views on this subject have been disputed by the U.S. Treasury Department).[39] Thus, the United States acknowledged that energy was at the center of relations between the industrial countries and the developing world, and that a close link existed between oil and raw materials. At the request of Saudi Arabia, a fourth item was added to the agenda: international financial problems.

The stage was set for the resumption of the Conference on International Economic Cooperation (CIEC). The membership was raised

---

36 Bernard Peyre, "La Conference Nord-Sud," *Défense Nationale*, March 1975, p. 45.

37 American Embassy, *USA Documents* (Paris), no. 2437.

38 American Embassy, *USA Documents* (Paris), no. 2448.

39 Dan Morgan, "U.S. Aides Split on Raw Material Trade," *International Herald Tribune*, February 18, 1976, p. 7.

to twenty-seven. The developing countries numbered nineteen, including seven members from OPEC. Representing the industrialized countries were Australia, Canada, the European Economic Community, Japan, Spain, Switzerland, Sweden, and the United States.

It proved difficult to reconcile the procedural and policy views of the EEC member states. British Foreign Minister James Callaghan claimed a separate seat for his country on the grounds of Britain's new position as an oil producer and of the absence of a common EEC energy policy. At the summit meeting of the EEC in Rome on December 3, 1975, Britain was forced to relax its stand, but obtained the assurance that some sort of minimum safeguard price for oil would be established.

World economic talks started in Paris on December 16, 1975, with each of the four commissions headed by cochairmen who were representatives of the industrial and the developing countries. The European Community acted as cochairman of the development and financial questions commissions, and Japan of the raw-materials commission. Significantly, the energy commission was cochaired by the United States and Saudi Arabia.

At their meeting in July 1976, the participating countries were unable to agree on the priority agenda of the four commissions. The developing countries insisted that the industrialized countries should make firm commitments on two issues essential in their view to the establishment of a new international economic order: the stabilization of the purchasing power of export receipts, and the drastic reduction of external debts.

There were other bones of contention. In the field of energy, the participants presented their forecasts on production, consumption, and alternative resources, but OPEC seemed likely to claim an oil price indexed to the prices of imports from industrialized countries. In the field of raw materials, a consensus might be reached over a period of time on several commodity agreements, if the United States took a favorable stand. On finance, it appeared unlikely that the industrialized countries would agree to discuss the reform of the international monetary system.[40]

With its broader scope, the conference may have achieved a better balance than if it had been exclusively centered on oil. The various groups of countries have claims allowing for give and take, if they regard negotiations not as a zero-sum game in which a gain for one side must be at the expense of the other, but as a positive-sum game in which each player can expect to be better off than before.

---

[40] Peyre, "Conference Nord-Sud," pp. 50-53.

European countries as a group will probably keep a low profile in view of their lack of common ideas on many energy questions, their monetary troubles, and the mediocre prospects for political union. The conference will consist largely of a United States-OPEC dialogue, if not a United States-South dialogue.[41]

At stake in the New International Economic Order (NIEO) according to Pierre Desprairies is a power struggle for "the sharing between two groups of countries of the growth rate of the world economy." The industrialized countries cannot keep for themselves the value added to raw material by the transformation processes. They must accept the industrialization of the producers of fuel and raw material, and therefore the transfer of technology. Industrialized countries are already engaging in primary activities, such as the production of substitute energy, and developing countries in secondary and tertiary activities. Both trends lead to less international division of labor.[42]

Prospects and Options for Western Europe

The claims for a New International Economic Order and the launching of the Conference on International Economic Cooperation point to the necessity for the industrial countries to define their economic future in geostrategical terms beyond the sole problem of energy, important as it is. In this respect the medium-term economic prospects for Western Europe are far from bright, and the political options are limited.

Economic Prospects. The starting point for an assessment of Western Europe's economic prospects in the medium term—the time horizon chosen being 1985—is a realization that raw materials in the broad sense accounted for 58 percent of total imports in 1972.

Can this high proportion eventually be reduced? The possibilities for substitution and recycling are limited for nonfuel-nonfood commodities which, according to Table 14 account for 18.8 percent of imports.

The EEC could certainly over a period of time make better use of arable land and reshape agricultural output so as to reduce imports of food products and develop agricultural exports to its suppliers of fuel and industrial raw materials. But the United States and Canada are

[41] Jacques Vernant, personal communication to the author.

[42] Pierre Desprairies, "Le Futur ordre pétrolier: économie de marché ou justice internationale," *Revue de l'énergie*, February 1976, pp. 79-81.

Table 14

EEC IMPORTS OF COMMODITIES FROM NON-EEC COUNTRIES
(percent)

Food, tobacco, et cetera	19.4
Other agricultural commodities	9.3
Ores and minerals	4.8
Fuel	19.8
Nonferrous metals	4.7
	58.0
Nonraw materials	42.0
All non-EEC country imports	100.0

Source: Hans Michaelis, *Memorandum über eine europäische Rohstoffpolitik* [Memorandum on a European raw materials policy] 1975, p. 20 (rounded figures), as quoted in Wolfgang Hager, "Europe's Economic Security," *The Atlantic Papers*, March 1975, p. 42.

in a dominant position as world grain exporters and have the capacity to increase output in relation to demand. Therefore a shift in EEC common agricultural policy would require negotiations with the United States. At the same time, imports of fuel were expected by the OECD and the EEC to remain at approximately the same level of volume in 1985 as in 1972, under certain assumptions of development of domestic primary sources of energy (see Chapter 1). Both organizations expected Western Europe's dependence on external energy to fall from 65 percent in 1972 to 40 percent in 1985. However, today the EEC commission considers this evaluation too optimistic. Thus external dependence will still be in the 50 percent range in ten years time (see Chapter 4).

In the present disarray of international trade and monetary relations, enormous uncertainties cloud the future world price of energy.[43] According to Neil Jacoby, the conditions that would lead OPEC to lower the price of its oil are the following: "The ending of rapid price inflation, a settlement of the Middle East conflict, and strong action by the United States to move toward energy independence."[44] Prospects for the implementation of the last two conditions, discussed

[43] Willrich, "Energy Independence," pp. 60-61.

[44] Neil H. Jacoby, "Oil and the Future: Economic Consequences of the Energy Revolution," *Journal of Energy and Development*, Autumn 1975, p. 50.

above (Chapter 5), are not encouraging. In the area of fighting inflation, West Germany has the best record, while other European countries are reluctant to follow even mildly deflationary policies for fear of social upheavals.[45] The Mediterranean countries more than the others suffer the dual effects of inflation and deflation, as they are struck by reduction of money brought in by their migrant workers and by recession in the tourist industry. An industrial area with scarce raw materials, Western Europe is at a disadvantage in relation to vast continents well endowed with natural resources, like the United States and the Soviet Union. The latter can bear the brunt of an increase in relative prices of commodities, the former has difficulties in balancing its trade.

What are the prospects for European exports? As explained in Chapter 2, a new international division of labor is inevitable. Some energy-intensive industries, such as petrochemicals, fertilizers, and the smelting of nonferrous metals, will be moved to countries where primary energy is cheap. Producers of energy and raw materials will engage in the downstream activities of their natural resources. European industries will face the problems of negotiating the sharing of the value added to raw material by manufacture, together with the transfer of technology. Non-oil-producing developing countries will continue to bank on labor-intensive industries.

The persistent fragmentation of the European market will operate in favor of sophisticated U.S. industries such as those in the computer, aircraft, and nuclear reactor fields. Thus Europe may be squeezed between the industrialization of the oil-producing and non-oil-producing developing countries and the marketing and technological efficiency of the United States. On the positive side, Europeans will benefit from sales of equipment to the developing countries but will be exposed to U.S. and Japanese competition.

Faced with the prospect of serious imbalance in foreign trade, European countries may resort to various forms of protectionism, such as the establishment of import quotas, voluntary export restrictions, and cartel and market-sharing agreements. They may also choose to engage in competitive devaluations. Such trends are already apparent in the economic policies of Italy and Britain, and may soon extend to France, while Spain and Portugal are in a precarious monetary situation. The current monetary debacle in Western Europe has been to a certain extent the aftermath of the oil crisis. Unfortunately the stabili-

[45] The percentage rise in consumer prices at an annual rate in 1976 was 10.8 percent for OECD Europe as against 5.8 percent in the United States and 4.5 percent in Germany (*OECD Press Release*, February 7, 1977).

zation of weak currencies is improbable. The recession of 1974–75 reduced industrial and domestic demand for oil and corporate demand for bank credit. As economic recovery gets under way, both demands will rise, and deficits in the balance of current payments in the weaker European countries will tend to increase.

To summarize, European countries will be losing on the trade side and on the production side. They will have to foot a large part of the bill of the New International Economic Order, to the detriment of the standard of living of their populations.

Political Prospects. Two areas must be considered: internal relations and external options. At their December 1974 meeting, EEC heads of state and of government invited the Belgian prime minister, Leo Tindemans, to report on the purpose and means of implementation of the "European Union" previously set forth as an undefined goal.

Prime Minister Tindemans, in his subsequent report (December 1975), stated that in the course of his research he was struck by the general feeling of Europe's vulnerability and impotence. He stressed that, without the political consensus of member states, institutions of the European Union would always lack soul and strength, and he assumed that EEC states were prepared to hold an election in 1978 of the European Assembly by universal suffrage. Finally, he recommended the creation of a common foreign policy on the basis of majority vote, and the consideration of a common defense policy, since defense commands foreign policy.[46]

The European Council (an informal version of previous summits), in its meeting of July 1976, reached a decision on the number of seats which European countries will have in the projected European parliament, whose members will be directly elected. But in September the ministers discarded Tindemans's suggestions for strengthening the parliament, and for substituting majority votes for unanimous votes by the Council in the field of foreign policy. Consequently, the chances for the creation over the next decade of a European union with any attributes of power are practically nonexistent.

The impotence of the European Community diagnosed by Tindemans has been enhanced by the situation of some southern European countries. In Spain, the crisis of succession after General Franco's death has been less serious than anticipated, but the whole political climate remains, to say the least, uncertain. In Portugal the return to a pluralist political system is far from assured. In Italy,

[46] Leo Tindemans, *The European Union*, Report to the European Council (Brussels, December 1975).

the Christian-Democratic government can survive only with the support of the Communist party. In France, success of the Socialist-Communist coalition at the next elections is a distinct possibility. These political situations may impinge on the external options of the two countries.

The Western European countries, with many nuances ranging from full commitment to armed neutrality, have relied since 1949 on the Atlantic Alliance, and have not set up a military organization of their own. Exposed to a potential threat from the Soviet Union, Western Europe is dependent upon the United States for its military security. Exposed to a reduction in oil deliveries, it is dependent upon the Middle East for its economic security. A sudden awareness of the lasting gap between their needs and resources in energy, and of their consequent economic insecurity, has not led Europeans to deviate from their passive attitude in regard to defense. Their naval forces in the Mediterranean and the Indian Ocean are negligible compared with those of the United States and the Soviet Union. Safe oil transportation to Europe is linked to U.S. naval presence. Therefore a "Mediterranean policy" for the European Community consisting in privileged trade relationships can only be implemented in the framework of the Atlantic option.

A drift from the U.S. umbrella to a Soviet protectorate is conceivable. Such a trend might emerge through a combination of factors, such as the advance of neo-Marxist ideology and of its supporting political parties, the disbelief in any sort of Soviet threat, and the mishandling of Euro-American relations in economics and defense, leading to a gradual weakening of the Atlantic Alliance. It could culminate in "Finlandization," the yielding of European governments to Soviet pressures and promises both in domestic and in foreign affairs. Alternatively in some countries it could lead to a right-wing reaction which in turn might provoke a rebellion of trade unions with the germs of a civil war. Energy considerations would play a part in these developments only to the extent that the increase in oil prices adds hardship to a crisis situation.

In the United States, the control of one or of several Western European governments by Communist parties is perceived as the major danger, the main goal of U.S. diplomacy being the containment of the Soviet Union along the present line dividing Europe into two zones of influence.

In reviewing the economic and political prospects of Western Europe in a world setting, Raymond Vernon commented on the distribution of power: "No single pattern can be regarded as highly probable in a situation so loaded with potential crisis."[47] Predominant among the causes of present instability and among the future uncertainties is the conflict between the growth of economic independence and the development of nationalism. To quote Mason Willrich again, "In a world of politically independent nation states, too much economic interdependency may lead to insecurity in particular nations and thus to instability in the international system as a whole."[48]

Europeans have not endeavored to master this potentially dangerous contradiction. At the level of the citizen, years of affluence have raised endless expectations, so that Europeans, living above their means, are not making substantial sacrifices for energy conservation. At the regional level, no effective integration has been achieved towards reducing the Community's external dependence in the areas of the economy and of defense. For the time being, the countries of Western Europe are following more or less willingly U.S. leadership in energy, in spite of some conflicting economic interests. But in the absence of a common political will, policy options of the larger European countries may diverge in the medium term more strongly than they have since 1973.

At their Paris meeting in 1972, European heads of state and of government expressed the view that Europe must be able to make its voice heard in world affairs and to make its mark as a distinct entity. In an official policy statement, Secretary of State Kissinger observed: "The period in which Europe was attempting to define itself is at least temporarily over."[49] In the views of Henry Simonet, a member of the EEC Commission, "European identity presents itself in the form of a question mark."[50] To put it bluntly, the energy crisis and its aftermath have revealed the absence of a European identity.

---

47 Raymond Vernon, "The Distribution of Power," *Daedalus*, Fall 1975, p. 256.

48 Mason Willrich, *Energy and World Politics* (New York: The Free Press, 1975), pp. 2, 26.

49 State Department official non-verbatim summary of remarks made by Secretary of State Henry Kissinger at a meeting of U.S. ambassadors in Europe in London, December 1975, as reported in the *International Herald Tribune*, April 12, 1976.

50 Henry Simonet, "The Energy Debate and Atlantic Relations," *NATO Review* (Brussels, February 1976), p. 6.

APPENDIX

Units of Measurement

Since this study deals primarily with energy in Europe, most measurements have been expressed in European terms. For readers unfamiliar with these units, the author has provided the following definitions and conversion factors.

APPROXIMATE CALORIFIC EQUIVALENTS

One million metric tons of oil equals approximately:

1.5	million tons of coal
1.167	million cubic meters of natural gas
41.2	billion cubic feet of natural gas
12.0	billion kilowatt-hours
10.000	teracalories

Heat units and other fuels in terms of million metric tons of oil:

1 million tons of coal	0.67
1 billion cubic meters of natural gas	0.86
1 billion cubic feet of natural gas	0.24
10 billion kilowatt-hours	0.82
10.000 teracalories	1.00

Source: *BP Statistical Review of the World Industry 1975.*

ABBREVIATIONS

bl/d	barrel per day	m^3	cubic meter
cf	cubic feet	MWe	megawatts
cf/d	cubic feet per day	SWU	separative work unit
GWe	gigawatts	t.c.e.	tons of coal equivalent
m	million	t.o.e.	tons of oil equivalent

CONVERSION FACTORS FOR CRUDE OIL

Unit Converted From	Unit Converted to: Metric tons	Long tons	Short tons	Barrels
Metric tons	1.000	0.984	1.102	7.33
Long tons	1.016	1.000	1.120	7.45
Short tons	0.907	0.893	1.000	6.65
Barrels	0.136	0.134	0.150	1.00

Conversion of barrels/day to tons/year: 49.8

Conversion of tons/year to barrels/day: 0.0201